MIDDLE EASTERN CITIES

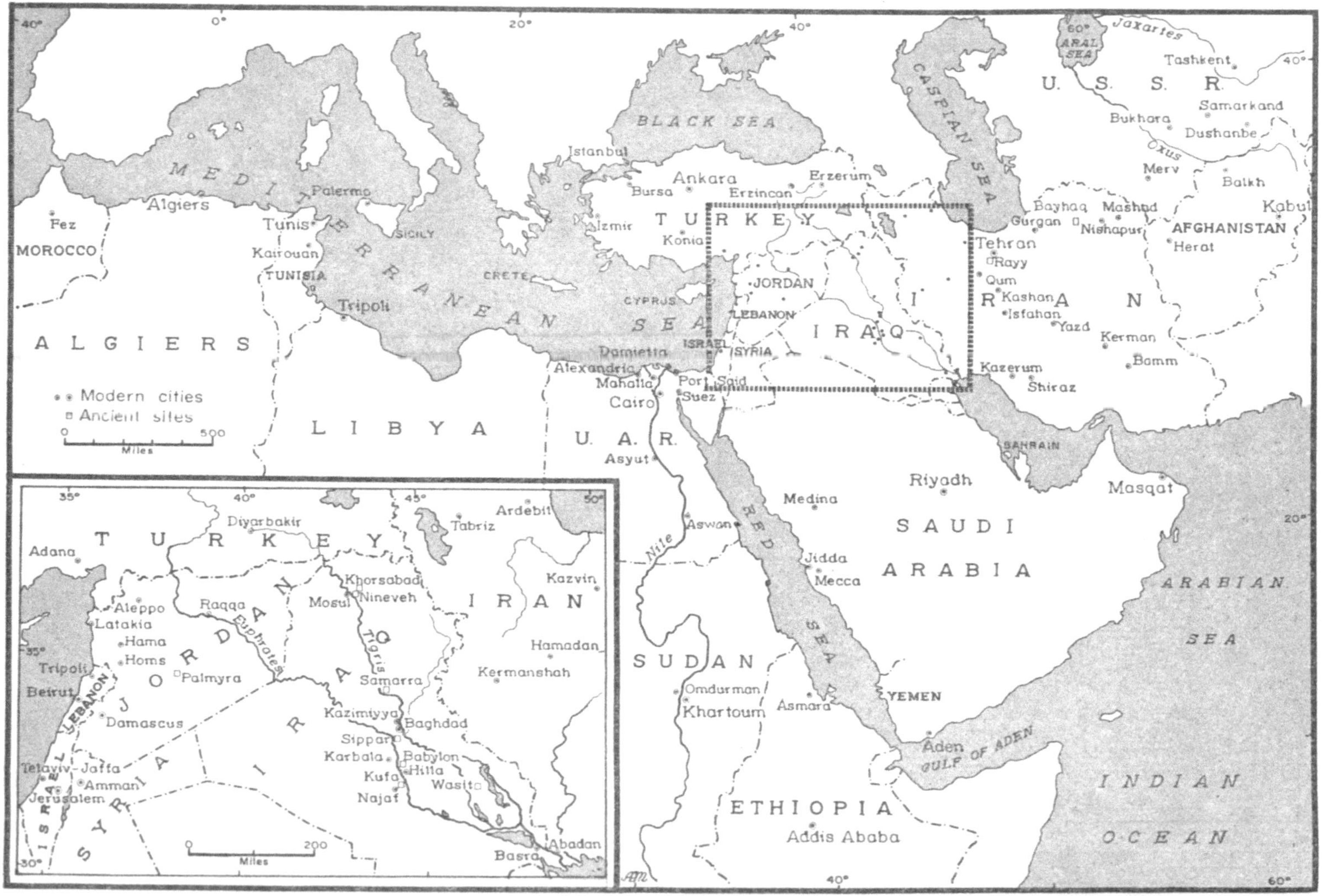

MEDITERRANEAN SEA
BLACK SEA
CASPIAN SEA
ARAL SEA
RED SEA
ARABIAN SEA
INDIAN OCEAN
GULF OF ADEN
Jaxartes
Oxus
Nile
U. S. S. R.
Tashkent
Samarkand
Bukhara
Dushanbe
Merv
Balkh
Kabul
AFGHANISTAN
Herat
Bayhaq
Mashhad
Nishapur
Gurgan
Tehran
Rayy
Qum
Kashan
Isfahan
Yazd
Kerman
Bamm
Kazerum
Shiraz
IRAN
TURKEY
Istanbul
Bursa
Ankara
Erzincan
Erzerum
Izmir
Konia
JORDAN
LEBANON
SYRIA
ISRAEL
IRAQ
CYPRUS
CRETE
SICILY
Palermo
Algiers
Fez
MOROCCO
Tunis
Kairouan
TUNISIA
Tripoli
ALGIERS
LIBYA
Damietta
Alexandria
Mahalla
Port Said
Suez
Cairo
U. A. R.
Asyut
Aswan
SUDAN
Omdurman
Khartoum
ETHIOPIA
Addis Ababa
Asmara
Medina
Jidda
Mecca
Riyadh
SAUDI ARABIA
BAHRAIN
Masqat
YEMEN
Aden
Modern cities
Ancient sites
0 500
Miles
40° 0° 20° 40° 60° 20° 40°
TURKEY
Adana
Diyarbakir
Tabriz
Ardebil
Kazvin
IRAN
Hamadan
Kermanshah
Khorsabad
Nineveh
Mosul
Raqqa
Euphrates
Tigris
Aleppo
Latakia
Hama
Homs
Palmyra
Tripoli
Beirut
LEBANON
Damascus
JORDAN
IRAQ
Samarra
Kazimiyya
Baghdad
Sippar
Karbala
Babylon
Hilla
Kufa
Wasit
Najaf
Basra
Abadan
Telaviv-Jaffa
Amman
Jerusalem
ISRAEL
SYRIA
SAUDI ARABIA
0 200
Miles
35° 40° 45° 50° 30°

Middle Eastern Cities

A Symposium on Ancient, Islamic, and Contemporary Middle Eastern Urbanism

edited by
IRA M. LAPIDUS

University of California Press · Berkeley and Los Angeles 1969

University of California Press
Berkeley and Los Angeles, California

University of California Press, Ltd.
London, England

Library of Congress Catalog Card Number: 72-81939

PREFACE

From the beginning of recorded history Middle Eastern cities and civilization have been one and the same. In ancient and medieval times, cities were centers of social and political organization and cultural creativity. In modern times, they are central to the processes of economic development and modernization which profoundly alter Middle Eastern society.

To review our knowledge of this important subject, to bring together the data of different researches and disciplines for evaluation and interpretation, and to chart the problems of future research, the Committee for Middle Eastern Studies of the University of California at Berkeley, with the Department of City and Regional Planning and the Center for Planning and Development Research, called a conference on Middle Eastern urbanism. This conference was held on October 27–29, 1966.

The conferees represented scholars working on different periods and regions, working with different data and methods of study, differing even in the definition of their research and intellectual concerns. Despite the varied orientations of the participants, the conference came to focus on two themes. The first is the relation of the city to its surrounding environment, conceived by some in political or cultural terms, by others in geographical, ecological or economic terms. Dealing with different periods, evidence and methods of analysis, the various papers explore the relationships between urban and rural areas. Questions are raised about the structural similarities between towns and villages. To what extent do their populations differ in occupations, in social organization, in attitudes and values? Various papers also examine the interactions between city and country, such as population movements and the transmission of cultural influences. Further, the conferees consider the implications of these relationships for the character of Middle Eastern societies.

The second major theme is the study of internal organization and the forces which influence the formation of city societies and the changes which these societies undergo. Group and class divisions within populations, and, alternatively, the bases of social unity are examined. Kinship, quarters, religion and government are subjects of discussion. Professor Oppenheim considers ancient city municipal organization; Professor Grabar, changing forms of Muslim religious expression; Professor Lapidus, religious communities and their social functions; Professor Goitein, business partnership and properties; Professor Issawi, demographic changes; Professors Gulick and Abu-Lughod, social change, new populations, and their assimilation to common cultures. Weaving these themes into the larger contexts defined by their interests and methods, the conferees attempt to illuminate the meaning of any single aspect of society by its ramified connections to the whole.

To follow these discussions, the reader should be aware of the format of the conference and its presentation in this volume. The conferees first met to discuss drafts of their respective contributions. Following formal comment by the discussants, the papers were open to general discussion, and subsequently they were revised for publication in this volume. Here they are grouped according to period, briefly introduced by the editor who attempts to indicate the general historical background and call the reader's attention to the main themes of each paper as they bear upon the others. A concluding essay by Professor Adams considers some of the main substantive and methodological themes which have emerged from the conference, analyzing the major continuities and disjunctions in the experience of Middle Eastern cities and the need for theoretical formulations in urban studies. Otherwise, no effort has been made to impose any formal unity on the presentation of the papers.

Appended to each paper is a partial record of the conference discussions. From the extensive conversations of the symposium, only those remarks which reflect differences of perspective and methodology, differences of opinion and judgment, and new information have been edited for inclusion in the final volume. Much discussion and many observations which were invaluable to the participants, and much of the process of questioning and response which helped clarify ideas and expand understanding, have been omitted since the conclusions are embodied in the revised papers. The selection of materials in no way reflects the relative contributions of the participants, many of whom made invaluable observations which, having fulfilled their purpose, are no longer pertinent in a printed statement.

Thus, the discussions are not presented verbatim, but are edited to bring relevant remarks together with those essays for which they seem most pertinent. In the course of the meetings, discussion ranged freely over many points, and the editor's grouping of materials does not reflect the actual transactions of the meetings but his judgment as to what arrangement of materials would facilitate the reader's coming to grips with the main ideas expressed. The presentation of the book follows the logic of the papers. The book is not a history of the conference, but a presentation of the collective outcome of the participants' activities.

This conference would not have been possible without the generous initiative, cooperation and support of many people and institutions. The editor wishes to acknowledge the unseen, but essential, activities of the members of the Committee for Middle Eastern Studies who organized the conference: Professors William M. Brinner, Wolfram Eberhard, Jonas C. Greenfield, George Lenczowski, Laura Nader, John M. Smith, and Professor John W. Dyckman of the Center for Planning and Development Research. The conference is particularly indebted to Professor Lenczowski who originally conceived and planned for this meeting, and to Professors Brinner and Nader for their efforts in planning and making the myriad of necessary arrangements. The Institute for International Studies of the University of California and the Joint Committee on the Near and Middle East of the American Council of Learned Societies and the Social Science Research Council generously sponsored the meeting of the participants. Not least does the editor thank Miss Bonnie Schurr and Miss Sharron Brown for their invaluable help in editing the text, Lorna Price of the University of California Press for her tasteful judgement and painstaking efforts to prepare this book for publication, and the contributors and discussants for their generous cooperation in the presentation of this volume.

I. M. L.

CONTENTS

PART I. *The Ancient Middle Eastern City* xiii

Introduction to Part I 1

Mesopotamia—Land of Many Cities 3
A. L. OPPENHEIM

PART II. *The Traditional Muslim City* 19

Introduction to Part II 21

The Architecture of the Middle Eastern City 26
O. GRABAR

Muslim Cities and Islamic Societies 47
I. M. LAPIDUS

Cairo: An Islamic City in the Light of the Geniza Documents 80
S. D. GOITEIN

PART III. *Contemporary Middle Eastern Cities* 97

Introduction to Part III 99

Economic Change and Urbanization in the Middle East 102
C. ISSAWI

Village and City: Cultural Continuities in Twentieth Century Middle Eastern Cultures 122
J. GULICK

Varieties of Urban Experience: Contrast, Coexistence, and Coalescence in Cairo 159
J. ABU-LUGHOD

CONCLUSION 188
R. McC. ADAMS

INDEX 197

PARTICIPANTS IN THE CONFERENCE

CONTRIBUTORS

Janet Abu-Lughod
Associate Professor of Sociology, Northwestern University

Robert Adams
Professor of Anthropology, Oriental Institute, University of Chicago

S. D. Goitein
Professor of Arabic, University of Pennsylvania

Oleg Grabar
Professor of Near Eastern Art, University of Michigan

John Gulick
Professor of Anthropology, Chairman of the Department of Anthropology, University of North Carolina

Charles Issawi
Ragna Nurkse Professor of Economics, Columbia University

Ira M. Lapidus
Associate Professor of History, University of California, Berkeley

A. Leo Oppenheim
Professor of Assyriology, Oriental Institute, University of Chicago

DISCUSSANTS

William M. Brinner
Professor of Near Eastern Languages, Chairman of the Department of Near Eastern Languages, University of California, Berkeley

Robert A. Fernea
Associate Professor of Anthropology, Director, The Middle East Center, The University of Texas

Peter Geiser
Professor of Sociology, California State College, Hayward

Gustave E. von Grunebaum
Professor of History, Director, the Near East Center, University of California, Los Angeles

Anne D. Kilmer
Associate Professor of Assyriology, University of California, Berkeley

Jørgen Laessøe
Professor and Director of the Institute of Assyriology, University of Copenhagen, and Visiting Professor of Assyriology, University of California, Berkeley

George Lenczowski
Professor of Political Science, University of California, Berkeley
Clement H. Moore
Assistant Professor of Political Science, University of California, Berkeley
Laura Nader
Associate Professor of Anthropology, University of California, Berkeley
Nadav Safran
Associate Professor of Government, Harvard University

PART I

The Ancient Middle Eastern City

Introduction to Part I

Mesopotamian cities were mankind's first cities. From relatively small village communities, with their low level of culture and lack of occupational specialization or social complexity, emerged the literate culture, the monumental arts, the highly developed economies and societies of Babylonian cities. Agriculture was supplemented by a complex trade, simple society was superseded by a highly stratified society, and rudimentary governing councils were replaced by developed monarchies and temple communities. From Babylonia of the fourth and third millennia, the impetus towards city growth spread to far-flung regions.

In his paper, Professor Oppenheim considers some of the essential qualities of these most ancient cities. He examines the ancient Mesopotamian city in its relation to the surrounding countryside, appraises its size and physical form, and analyzes the implications of observed physical structures for social organization. Reviewing the role of temple and palace, or god and king, in the governance of the ancient cities, Professor Oppenheim weighs the extent to which an autonomous urban community had emerged in the ancient world.

To explore this central theme in more depth, Professor Oppenheim presents the latest findings on the study of one particular city, Sippar, known from documents dating from the nineteenth to the sixteenth centuries B.C. He reflects further upon the organization of families and classes, the nature of local governing institutions, the selection of "mayors" and their relations to royal authority, and the meaning of citizenship. Professor Oppenheim completes his discussion of the polity and society of Sippar in the Old Babylonian period by a review of the city's economy, stressing trading activities and a surprisingly commercialized agriculture.

Throughout his comprehensive, though succinct paper, Professor Oppenheim touches on important themes to which the conference often returned. First are questions concerning the city's relationship to rural populations and sources of food, and its roles in trade and in empire-wide politics. Second are questions of internal organization—the meaning of citizenship and local freedoms, and the influence of local notables. Comparisons and contrasts emerge in the subsequent discussions of both traditional Islamic and modern Middle Eastern cities, and are reviewed in the concluding essay by Professor Adams.

A. L. OPPENHEIM

Mesopotamia—Land of Many Cities

The fame of the two largest cities of Mesopotamia, Babylon and Nineveh, was based, until a century or two ago, mainly on the Old Testament and on Herodotus. These works speak with censure as well as admiration of the size, the riches, and the famous buildings of these capitals. When the writing systems of Mesopotamia were deciphered and the language of its documents understood, they yielded a staggering amount of new information to Sumerologists and Assyriologists. The names of many more capitals, other cities, and towns became known. This information aided the archeologists in finding and identifying the ruins of the many ancient cities that they have been excavating ever since. Apart from the famous sites of Babylon and Nineveh, a considerable number of large cities were discovered, such as Nippur, Uruk, Ur in Babylonia proper, Assur, Calah, and Khorsabad in Assyria, as well as many provincial towns and even smaller fortified settlements. This substantial body of information—which covers much of the area of the ancient Near East (today's Iraq, Syria, Turkey and Israel, apart from Egypt) during a period of nearly three millennia—enables us to speak with considerable assurance about the size of such cities, their fortifications, the layout of the streets, the position of the prominent public buildings, the relationship between topography and city plans, and so on. Of course, a city consists not only of buildings, streets, and walls but also of people, fused together by specific social experiences and dominated by characteristic social attitudes in their relationship to each other and to the outside world. A city has a history and makes history; a city is a body politic which evolves its own rules and its own spirit.

The civilization of Mesopotamia has left us an abundance of records but rather little direct evidence of city life, city administration and politics. The background of tensions and conflicts is not easily detected in the stereotyped and formalized phraseology of the Mesopotamian scribes when they refer to historical events in which these cities played a crucial role. Moreover, what we learn from material of this nature is bound to be one-sided and restricted.

Urbanization occurs quite early in the third millennium B.C. in Southern Mesopotamia. The processes which antedate this event as well as the first half-millennium or so of that development are totally beyond our reach; it may well be that only speculation will ever fill this gap. Cities in Southern Mesopotamia appear densely along water courses—even within sight of one another. Their names are neither Sumerian nor Akkadian, indicating that these names were chosen by an ethnic group or groups which spoke neither of these languages. As important as the density of cities in the south of Babylonia is the fact that urbanization did *not* take place in the large arc of land extending from the Persian Gulf along the twin rivers of the Euphrates and the Tigris, across Syria, and southward along the littoral of the Mediterranean Sea down to and probably including Egypt. It is true that cities arose here and there in that region, but only around royal residences, as trading settlements on the coast or on rivers, or at wells and in certain holy places. Nowhere do we find the agglomeration of walled towns which existed so early in Southern Mesopotamia.

Even though the growth of urbanization was important in Mesopotamia and had fateful consequences for the history of the entire region, we should not forget that all these cities were embedded in an expanse of open country with a socio-political mood and an historical development of its own. Who lived in what was called "country" (Sum. m a. d a, Akk. *mātu*) and how, we are not likely ever to know fully. There is very little direct documentation and only rarely is there any archeological evidence for the peoples who inhabited that large section of Mesopotamia. Besides villages (sometimes called *kapru* or *edurû*) there were larger and older towns (Akk. *ālu*; Sum. u r u) , fortified garrisons beside ephemeral settlements consisting of encampments of tents and reed huts, and police outposts. Densely populated stretches along newly dug canals bordered abandoned fields ruined by salinization and drifting sand, and pasture grounds alternated with low-lying stretches used for quick crops suited to the moisture left there by receding floods. But we have no way of learning about the density of the population in such disparate habitats, or about the economic situation and social setup of these people, let alone their political role or potential. The legal and administrative tablets and the royal inscriptions—our main sources of information—are primarily concerned with city life and city power. Only occasional allusions in private letters and in literary texts provide a glimpse of that side of Mesopotamian social and political life of which even the scribes most probably had only a rather vague notion. The open country was, at all times and in all regions, as important in the political structure of Mesopotamia as the city and the king, though its role in the evolution of Mesopotamian cultural and social traditions remains to be investigated. For the present purpose it should be sufficient to point out one specific facet: the antagonism of residents of the open country to cities and their inhabitants. From the emergence of the first city-states to the conquest of our region by the Arabs—and even later on—Mesopotamian history could be written largely in terms of pro- and anti-urban policies, or, from another point of view, in terms of genuine and artificial urbanization. In the latter situation, the royal policy of deliberately creating new cities (mostly in non-urbanized regions) had several different purposes: to create administrative foci for the extraction of taxes and services from the recalcitrant and unstable inhabitants of the plains and the mountain valleys, to ensure dominion over newly conquered territory, to levy dues on caravans, or to counterbalance the political power of the old cities.

We should not neglect the important social fact that city and country dwellers were in steady, osmotic interaction. On the one hand, refugees from debts and other obligations left the city and presumably joined segments of the rural population which were less than sedentary. On the other hand, the city, with its opportunities for an anonymous existence in the heterogeneity of the urban agglomeration, must have attracted country folk who saw there the possibility of selling their labor or whatever skills they had. Diplomatic relations of some sort also must have existed between country and city since overland trade normally depended on arrangements made with the tribes and nomads controlling the passage of the caravans or boats.

Of these cities, Babylon was the largest. In the middle of the first millennium B.C. it covered 2,500 acres. Next in size was Nineveh with 1,850 acres, the last capital of the Assyrian empire. Uruk, the biblical Erech, occupied 1,100 acres. Other cities were much smaller, such as Dūr-Šarrukīn with 600 acres, and Calah, likewise a capital of Assyria, with 800 acres. These last two were royal foundations while the old and holy city of Assur, the mother city of Assyria on a cliff overlooking the Tigris, extended over only 150 acres. For comparison, Athens at the time of Themistocles, encompassing 550 acres, was considered unusually large and populous. Thus we see that Babylon and Nineveh well deserved their fame as large urban agglomerations.

The typical Mespotamian city—and this we know from early Sumerian literary texts—consisted of three parts. First, there was the city proper (often called, in Akkadian, the inner city, *libbi āli*); then the "outer city" (u r u. b a r. r a), the suburb, for which we have quite a number of designations later on, in Akkadian; and third, the k a r, the harbor section. The city contained the palace of the ruler, the temples of the city's gods, private dwellings arranged along small, narrow, often crooked or dead-end streets, and a few wider streets mostly near the gates. It was divided into several quarters, each of which seems to have had its own gate through the walls surrounding the entire complex. The suburb also contained houses but mainly consisted of fields, date groves, and cattle folds which provided the citizens with food and certain raw materials. Fortified outposts seem to have guarded this green belt around the walls, at least in the first millennium. The harbor section was the center of the commercial activities of overland trade, and enjoyed administrative independence as well as a special legal status. Foreign traders seem to have lived there, probably provided for by what the texts call the "innkeeper of the harbor." Though it may well be that not all cities conformed to this threefold articulation, the arrangement invites speculation concerning underlying economic and ideological conditions. Special situations, such as those created by the crossing of trade routes, or due to strategic circumstances, have doubtlessly influenced this pattern. I am inclined to see the "suburb" as that corona of fields and gardens watered by the river or canal on which lay the city. The suburb—the corona—provided the economic foundation of the city whose citizens owned the fields or who were absentee landlords. This situation underlay the autarchic economic outlook that created the rigid separation between city and harbor. The foreign traders living on the pier transacted the kind of business which, as a rule, served palace and temple rather than the farmers, because they imported such materials as metal, precious stones, spices, perfumes and timber.

The fact that the city harbored both the palace of the ruler and the temple of the god must not be forgotten in a discussion of the Mesopotamian city. Both palace and temple operated basically as "internal-circulation organizations" centered in the person of the lord of the manor, deity or king. Their relationships to each other and to the settlement as such are very difficult to ascertain. The temple—where the god's image resided, the presence of which assured the city's well-being—does not seem to have exercised any appreciable political pressure on the city itself. It even seems at times that the poorer citizens benefited from the economic resources of the sanctuary while the rich citizens utilized the wealth of the sanctuary in their own overland trade ventures.

Relations between palace and city were quite different. The king's need for taxes and services (military and corvée duties) was by its nature contrary to the interests of the city and especially to those of the richer and hence more influential citizens. Although this conflict must have become rather acute at the end of the third millennium when the kings embarked on a far-reaching policy of conquest and imperial aspiration (typically Sargon of Akkad—about 2334–2279 B.C.), we have little knowledge about the details. Only by accident do we hear that about four hundred years later the citizens of Nippur (in central Babylonia) were officially exempted from military service and from paying tribute in gold and silver to the king of Isin. Several centuries later, the privileges of the oldest and largest cities—Nippur, Babylon and Sippar—were said to be under special divine protection. A religious symbol placed at the city's gate, probably a kind of standard (*kidinnu*), proclaimed its special legal status. First millennium literary texts reveal that the inhabitants of certain cities enjoyed specific legal privileges: for example, the king could not impose fines or imprisonment on them; nor was he permitted to dismiss without good reason their legal claims. Apart from freedom from corvée duties, the citizens were also protected against seizure of their plowing cattle by the king and against taxation imposed on their flocks. Such privileges were restricted to native-born citizens, but in a letter written to King Assurbanipal by the citizens of Babylon, we find the startling assertion that even a dog becomes free and privileged when he enters their city. Though such a statement reminds one forcibly of the medieval saying that the city's air makes those who breathe it free, it is more realistic to assume that in Mesopotamia the enforcement of all city privileges depended very much on the power situation, the political position of both the specific city and the ruling king. Still, the recurrent attempts of the kings, especially the Assyrian kings, to build their own capitals rather than to reside in the old and sacred cities (such as Assur) pointedly show that the cities had been winning their fight for freedom.

As for the spirit of the Mesopotamian city, the citizens completely accepted urban life. Literary texts reveal scarcely any resentment against the city, in marked contrast to certain passages of the Old Testament and also to later classical texts. Nor can one discover the vestiges of tribal organization that are so characteristic of the later Muslim cities of the region. The Mesopotamian city has no obvious ethnic or tribal articulations; it forms a primary social organization as a community of families of apparently equal status. It also existed as a corporation which could sell real estate, decide legal cases, write letters to kings, and receive messages from them.

More specific information about the city's political constitution, social structure, and institutions is contained in the Sippar Project of the Oriental Institute of the University of Chicago.[1] This project investigated the Babylonian city of Sippar by transliterating all legal and administrative tablets originating there. These tablets were datable to the Old Babylonian Period, the period of the dynasty of Hammurapi (the three centuries between 1894 and 1595 B.C.). They

[1] In 1962, the National Science Foundation gave a two-year grant for a study entitled *Urban Society, Economy and Demography in Ancient Northern Babylonia.* Under the supervision of Professor Robert M. Adams and myself, Dr. Rivkah Harris, a former student of mine, undertook the transliteration of the Sippar tablets. They are quite diversified in content and are, moreover, exactly datable, two essential conditions that are difficult to parallel in texts of this nature coming from any other Mesopotamian city. This material also happens to provide information of rather evenly distributed density over the period covered, which again is an important factor in an investigation of social and economic conditions extending over three centuries.

The research work was planned and carried through on the basis of an elaborate filing system. It now contains nearly 20,000 personal names; each person mentioned is listed not only alphabetically but also according to his office, function, or craft. Furthermore, the project has files with references to the size of all houses, fields, and gardens rented or sold; files of all geographical and topographical data; files of prices and rents paid for real estate, for slaves, for services and staples; and files of all technical terms used in legal and administrative contexts. It stands to reason that such an accumulation of social and economic data, most of them datable and hence helpful in tracing developments, can and should be used for many purposes. In fact these files are bound to reveal currently unsuspected problems, as well as information which may help solve them.

Mrs. Harris' manuscript, which will contain the bulk of the data, is now being prepared. To date interested readers may consult the following published articles:

Rivkah Harris, "The *Nadītu* Woman," *Studies Presented to A. Leo Oppenheim,* (Chicago, 1964), pp. 106–35.

——, "The *Nadītu* Laws of the Code of Hammurapi in Praxis," *Orientalia,* XXX (1961), 163–69.

——, "On the Process of Secularization under Hammurapi," *Journal of Cuneiform Studies,* XV (1961), 117–20.

——, "Biographical Notes on the *Nadītu* Women of Sippar," *Journal of Cuneiform Studies,* XVI (1962), 1–12.

——, "The Organization and Administration of the Cloister in Ancient Babylonia," *Journal of the Economic and Social History of the Orient,* VI (1963), 121–57.

——, "Old Babylonian Temple Loans," *Journal of Cuneiform Studies,* XIV (1960), 126–37.

A. Leo Oppenheim, "A New Look at the Structure of Mesopotamian Society," *Journal of the Economic and Social History of the Orient,* X (1967), 1–16.

number approximately 1,600—the largest corpus of documents known to have come from a single city. These texts bear mainly on the economy of the upper middle class, while high officials, the military, and the royal court are mentioned only incidentally, and the workings of the temple administration are rarely touched upon. The primary aim of the project was to trace the social structure of the city, the nature of its administration and its relationship to the capital, Babylon. We were also concerned with such interesting facts as average life span, size of family, social mobility, and sources of income.

For the following discussion I have selected different aspects of the Sippar material in order to delineate the complexities of Mesopotamian urbanism.

First, contracts between private parties were witnessed in Sippar as elsewhere in Babylonia by city and royal officials, priests and other professional men, and neighbors and relatives, all enumerated in the sequence of their rank with the most important witness listed first. Since these documents are dated, we can easily obtain significant information about the administrative structure of the city in any given year. The texts show that Sippar was administered on two levels that were strictly separated. On the lower level, the inhabitants of the city's neighborhoods or wards (*babtum*) regulated their local affairs, from sanitation to security, under an official called *ḫazannum* who was installed, apparently by the king, for irregular but considerable lengths of time. The second and higher level of administration was the one by which the city as a community—often called *ālum* (city) or *ālum u šībûtum* (city-and-eldermen)—related to the outside world and especially to the king. The designation *ālum u šībûtum* seems to refer to the entire free male population within the walls and to the heads of families of a certain social and economic status. Under specific circumstances these men acted as a body, but evidently the executive power was always in the hands of one official, usually the "overseer of the merchants" (*akil tamkārē*)—other designations of this position in certain periods notwithstanding. More important than such changes in the titulary is the fact that this official kept his office for only one year. He could hold it several times but never in successive years. Also, these men were all natives of Sippar, from very wealthy families. Through the many years which these documents record, the top functionaries came from a limited circle of persons.

This curious syndrome of facts demands an explanation. It seems contrary to the aims of a central administration to appoint such an official for the short term of one year. On the other hand, we have no precedent to suggest that the citizens of Sippar (or a certain segment of the population) elected their representative. However, Mesopotamian civilization employed an operational method of determining a sequence among peers that is outside human interference. This was the drawing of lots as it was practiced in the Old Babylonian period (in the south) to establish the sequence in which brothers were to select their inherited shares of the paternal estate; it was also practiced in Assyria when the *limmu*, the eponym official, was determined by lot at the beginning of each year (to which he then gave his name). Thus, it is not too farfetched to assume that the chief administrator of Sippar was selected by lot for an annual tenure and that the retiring official could not participate in the selection of his successor. This interpretation seems to fit the stated facts but fails to explain why so few persons were admitted to the selection. Conceivably there existed a council of eligibles consisting of the richest citizens or of the heads of the foremost families even though we have no evidence for such a council.

Even if certain problems connected with this institution remain unsolved, the very existence of these restrictions on the term of office of the head of the city government is remarkable. Equally important is the fact that power was wielded not on the basis of personal status (based on charisma, wealth, genealogy), but in rotation among peers who enjoyed what the Greeks called *isonomia*.

What circumstances could possibly have enforced such a system of restricted office tenure among persons of presumably equal wealth and influence but specific financial advantages? It is therefore necessary to search for the sources of these advantages. We know that the "mayor" of Sippar enjoyed the income from a royal field; hence, in some way he was in the king's pay. If this were the case, the following situation might be suggested as a possible background for the institution under discussion: there existed in an empire (like that ruled by the First Dynasty of Babylon) a basic conflict of interest between rich cities and the kings who were striving toward an effective control over these cities for taxing purposes, just as they tried to control the temples by means of their commissaries. Since cities in Mesopotamia had a tradition of urban autonomy, a compromise between these conflicting interests was necessary; perhaps the city's "mayor" was elected by lot from among those citizens whose personal financial status could guarantee the king the payment of the taxes imposed on the city. Thus, we seem to encounter in Old Babylonian Sippar a social mechanism somewhat akin to the "liturgy" of the Greek *polis* through which certain essential but costly public functions were conferred compulsorily on the rich citizens. I propose therefore that the reason for the rotating nature of the office of the "mayor" was that he was personally responsible to the central administration for the tax imposed on the city. In order to shift the burden equitably—and perhaps also to distribute in the same way the income and honors implied—lots were drawn every year under the supervision of the outgoing official.

These references to wealth and to taxes introduce the second point concerning the economy of Sippar. We know from a number of letters and legal texts of other cities that Sippar was an important emporium where river and overland routes intersected. To a large extent this was the consequence of its unique geographical position: no other city of its size can be found upstream along the Euphrates for a long distance. Also the famous tin road which brought this essential metal through the Diyala River Valley from Iran led through Sippar to the Euphrates route which linked Telmun, Ur, Emar, the Mediterranean coast and even Anatolia. Sippar, moreover, was a frontier town of considerable attraction for the tribesmen of the piedmont region, for the small farm communities along the rivers, for the nomads on the move at the fringes of the empire, and for the military outposts along its crucial border to the northwest. It is interesting to note that there is very little evidence concerning overland trade in the archives of Sippar which we are utilizing. The people who figure in our texts were wealthy indeed but they derived their income from agricultural holdings and they did not seem to participate actively in the overland trade for which Sippar was famous.

Two explanations can be proposed. First, the traders might have lived in Sippar, as elsewhere in Mesopotamia, in a special settlement outside the walls called *karum*, "harbor". But Scheil, who excavated in Sippar, as well as the Arabs, who, at the end of the last century, gathered there tens of thousands of tablets, never touched the ruins covering that *karum*. Second, there might have existed not only a spatial but also a social cleavage between "harbor" and "city," inasmuch as merchants became rich through trade but invested their money in land once they moved from the harbor into the city.

The emporium and frontier character of Sippar is underlined by the designation of certain of its quarters or suburbs, in which appear the names of two tribes well attested along the middle course of the Euphrates. Clearly, the section called Sippar-Yaḫruru must have grown out of an encampment of the West-Semitic speaking Yaḫruru tribe near the walls of the nuclear city. The same was undoubtedly the case with Sippar-Amnānum. They are a part of what our texts often call "Greater Sippar." This assembly of urban agglomerations (under separate administrations) contained a number of temples and other sanctuaries, streets and open spaces at the gates, and most likely no longer showed actual nomadic encampments.

The urbanization process was still in progress at the time of our texts. Only the peoples had changed: we read repeatedly about the tent encampments of the Kassites quite near Sippar. This is important because after the disappearance of the Kassite rule in Babylonia, more than half a millennium later, Kassite gods still had sanctuaries inside Sippar.

A curious bit of evidence seems to fit this situation. Several of our documents mention taverns and shops which appear to have been located near one another, even side by side, along the squares of two of these city quarters. I cannot tell whether or not other Babylonian cities of comparable importance and size possessed similar rows of shops and taverns inside the gate, but I suggest that this was not the case since the words used to denote them do not occur in rent and sale contracts from elsewhere in contemporary Babylonia. I venture to suggest that the goldsmiths, who are strangely enough the craftsmen most often represented in our archives, were located in these shops. There the nomads and farmers from the open countryside came when they visited the big city; there they bartered for food and golden trinkets with merchandise they brought to Sippar. But what merchandise, easily accessible to any tribesman, was so expensive that it could be exchanged for gold objects? Possibly, it was slaves and especially children who were brought to Sippar and then sold downstream. There are two reasons for this assumption. First, we actually have a number of sales documents from Sippar for small children coming from Subartu (roughly the region upstream along the Tigris and the mountain ranges); and second, we have a group of legal documents from the archives of rich persons in Sippar that deal with the division of property among heirs or with gift deeds made out to daughters in which as many as twenty and twenty-six slaves are listed, half of them males, half females. This contrasts with the paucity of slaves in the households of Sippar, a city in which only 17 percent of the known personal names refer to slaves. Moreover, only about half of all the available documents recording the transfer of private property through inheritance, gift, adoption and other transactions within the family—as well as litigations arising from such acts—refer to slaves at all. That the mentioned texts enumerate by names as many males as females shows convincingly that multiples of these numbers must have been in the possession of the persons who disposed of them. Clearly, here we are dealing with two entirely different socio-economic situations involving slaves: few slaves (mainly females for housework) in private households on one hand, and large numbers of them as stock-in-trade of slave dealers, on the other.

However, the main source of income of the people of Sippar whose activities are mirrored in the documents of our archives was not overland trade nor even slave trade, which is only rarely and indirectly attested. Instead, most Sippar families derived their income from nearby barley fields. They bought and sold such fields. They farmed them out or held them in tenancy, sharing the crop with the landlord. When the pertinent documents were examined, it was found that the archives illustrated an unexpected relationship between a big city and its agricultural hinterland—a relationship which I shall discuss here only as far as it bears on Mesopotamian urbanism.

In the Babylonia of the outgoing third millennium and the first third of the second millennium, barley was grown either on the large estates of the manorial organizations (temple and palace) by serfs and corvée workers, or on the small fields of free farmers who have left us many tenancy documents. The latter were constantly in debt and inevitably became something akin to sharecroppers in spite of royal attempts to relieve them of their burden of accumulated debts. From the Sippar documents, however, we learn that agriculture was commercialized. To grow barley was, for the urbanite, an undertaking which required none of his own physical labor at all. In certain instances, he did not even have to own fields. He was a type of entrepreneur who took advantage of the fact that growing of barley required but two periods of intensive field work which demanded a considerable labor force whenever the holdings exceeded the size manageable by the family. The two periods were the plowing and seeding, which in Mesopotamia was done in one completely mechanized operation (by means of a seeder-plow), and the harvesting. We have found a considerable number of documents from Sippar in which contractors in possession of the necessary equipment, of draft animals and trained workers, were hired for payment in silver to plow and to seed the fields with barley. The harvesting was done by workers who were hired either individually or by middlemen long before harvest time. These harvesters were well paid, often in advance, and were compelled by special ordinances to carry out their agreements, which prevented them from selling their services to the highest bidder. These two practices made it possible for the owner of a field to calculate quite accurately in advance the expenses involved in growing barley. This investment character of barley-growing in Sippar (where wheat is never mentioned and sesame very rarely) is furthermore illustrated by practices quite atypical for Mesopotamia: groups of evidently wealthy citizens pooled their fields and entrusted them to a manager (called *iššakku*); other investors even rented fields on a large scale either to have them worked by contractors and hired labor or managed by persons who took a share of the crop for their services. What was the economic reason for this particular development? The yield of these fields was in no way larger than elsewhere in Babylonia, nor were the agricultural possibilities of the Sippar region more advantageous with respect to irrigation, labor costs, etc. The incentive for such investment techniques can only have been an unusually good market for barley. This market may have been the city itself, meaning that segment of the urban population unable or unwilling to raise its own food, or it may have been the capital, Babylon. Since the Euphrates offered cheap and efficient transportation, and since we know from other sources that boats loaded with barley moved down the river quite frequently (even such distances as from Emar to Mari), a likely explanation is that the investors of Sippar grew barley for Babylon.

The Sippar archives also revealed, somewhat unexpectedly, the importance of a type of public granary *(našpakum)* capable of storing substantial amounts of barley. Every delivery and every withdrawal of barley was recorded in writing and the names of the acting officials were listed. It appears that the city as such was responsible for the contents of the granary and that therefore a strict system of written accounting of all transactions was used. This kind of bookkeeping is well known in Mesopotamia, although scholars have always thought it to be characteristic of the bureaucratic economics of temple and palace. Here it was applied, for reasons not yet evident, to account for barley belonging apparently not only to the citizens of Sippar but also to the royal administration. This is shown by the fact that as a rule, the documents are provided with the names of a scribe and of a high city official called either "Foreman of the Assembly" or "Overseer of the Merchants." While the latter clearly represented the city, the former must have been the palace representative.

The question arises whether the public granary owed its existence simply to lack of space for private granaries inside the crowded city or whether it represented an essential feature of the urban experience of Mesopotamian man. If the latter could be proven—e.g., by parallel institutions in other cities of the period and region—it would shed light on the history of urbanization in Mesopotamia. On the other hand, if a public granary should turn out to be characteristic only of Sippar, one would have to consider the possibility that the granary as such was a royal institution. We know that the granary in Sippar contained royal stores destined to feed the soldiers of the garrison, corvée workers repairing the city wall, and other royal employees. The sharing of the control of the granary by city and royal officials therefore poses a problem; it may have been a working agreement based on some form of compromise between these two powers of similar status.

One final point: in many Sippar contracts dealing with the sale of houses inside the city, we find houses or house lots described by the word *ezibtu* which means literally "left over, left behind." This word is found in legal texts from other Babylonian archives of the period, deeds which record the sale of fields. The designation "left over" may be explained as a term referring to small field lots "left over" when larger fields were sold. Certain land holdings were basically inalienable, and this seems to have led to the practice of the seller's retaining a small section of the original lot. Certain communal or administrative encumbrances were attached to land that was allotted as compensation for expected or actual services of office holders, and therefore it could not be sold to outsiders. However, when economic necessities compelled the holder to sell, a compromise solution was made; he retained a small token section of the land which enabled him to continue to enjoy his rights and privileges. The reason underlying such a practice in a crowded city is that the claim to citizenship and its privileges in a Mesopotamian city was based not solely on a person's having been born there of free parents, but also on the ownership of real estate inside the city's walls. These apparently were requirements for sharing in the economic, legal, social, and perhaps also religious facilities and privileges offered to the citizens. For such reasons, the owners of real estate refused to sell all of it and retained a small part—*ezibtu*.

This study must conclude with a *caveat*: For the entire duration of the project a reservation has lain in the back of my mind: is Sippar typical for Mesopotamia, for Babylonia, even for Northern Babylonia? It is difficult to tell since we do not know enough about other contemporary cities. There are a number of traits in the Sippar texts that seem to characterize the city as marginal. These are a set of Akkadian month names which do not recur elsewhere in Babylonia; cases of physical punishments stipulated in private contracts, a practice which clearly belongs to what I like to call the "Western-Barbarians"; and the use of certain words that point toward Mari and even Assur, to mention only the most outstanding instances. Possibly an intensive and perceptive analysis of the texts from another Mesopotamian city would divulge similar alien features—but nobody has ever attempted such an investigation. In short, we will have to content ourselves in the foreseeable future with material of the type the Sippar Project has produced if we intend to investigate Mesopotamian urbanism on the basis of primary evidence. This means that we will be able to scan only an accidental and rather narrow section of the population and will always be at the mercy of the peculiarities of a specific city and its specific situation. It is of course possible that a series of parallel investigations based on material from other cities will act as a corrective for such peculiarities and eventually enable us to control the variables. The first problem will always remain with us, as it does with any kind of research work restricted to documents of a dead civilization: written sources of a narrow topical range cannot reflect the workings of a complex civilization.

DISCUSSION

Professor Adams observed that there are additional reasons for which the example of Sippar may not be generalized to extend to other Babylonian cities: "Sippar was the northernmost of Babylonian cities, at the very edge of the steppes, occupying a special relationship to the nomadic tribes inhabiting that steppe, and also occupying a special role in whatever overland trade we can assume occurred between Anatolia and Babylonia, and possibly also between Iran and Babylonia. In this case, the emphasis on traders may be a peculiar artifact of Sippar's situation. Also it is interesting to note that Sippar never had a dynasty. Sippar cannot be compared with the major political centers of Babylonian antiquity.

"Furthermore, the manorial organizations mentioned by Professor Oppenheim may be more powerful, larger, and more numerous in certain kinds of urban settlements than in others. In Sippar, because of the special importance of trade, they were probably less important than, for example, in Babylon. Also, Sippar should not only be compared with Kish and Babylon but also with many small towns. Despite its economic importance, Sippar was a small town which occupied only ten or twelve hectares, and was similar to other small highly specialized communities, such as Harmal, which dotted the north Babylonian plain. I think we have to be prepared for a very sharp and unsuspected differentiation of town types and functions.

"Some of the apparent peculiarities of Sippar surely are a result of the available documentation. The tablets adequately reflect only the literate population and their reasons for writing. The absence of poor craftsmen, small traders, and so on, within the settlement may be an artifact of a documentation in which we cannot expect listings of people who were not engaged in economic relations with the upper-middle classes."

Professor Oppenheim acknowledged that we cannot say "what is typical and what is atypical so long as we know only Sippar. There was a great variety of urban types. The relationships between king, city, and palace were obviously variable. The size and location of the *oikos* of the king and that of the god are difficult to determine. We don't know how many people lived outside of the city, how much land outside of the city belonged to the temple, and how much to the king; nor do we know how many people belonged to the court, to the temple, or were citizens. There is no doubting the existence of important variations, and though Sippar may not be the last word, it is all we have."

Other questions about political and social organization prompted Professor Oppenheim to amplify some of his remarks. In connection with the discussion of local self-government, a question was raised about city assemblies and primitive city democracies. Professor Oppenheim observed that the existence of such assemblies in earliest times has been deduced from mention of an assembly of the gods in later texts, and he is uncertain to what extent we are entitled to assume their existence in fact: "Merchants were all on the same level and made decisions in committees, but nobody else could do that, neither slaves, nor people with restricted liberty, nor anyone of the court. The court was not free. People belonging to the royal court were slaves of the ruler. However, the mayor of Sippar was titled "chairman of the assembly." This has caused us much trouble because the assembly is never mentioned—only the title. I do not deny that in some cities of Mesopotamia there may have been assemblies, but in Sippar it was apparently only a dead title."

Dr. Milgrom asked about the existence of guilds in ancient Mesopotamian cities. Recent researches into traditional Islamic cities cast doubt on the existence of craft guilds in the early Islamic epoch, and Professor Oppenheim observed of ancient Mesopotamia that "there were no guilds in the sense of the late medieval guilds. In the second millennium, references are found concerning the head man of the carpenters and the head man of the iron workers. But these people, as far as I can see, were mostly connected with work in the temple. The craftsmen in the temple organization were under the direction of one man. This was not a guild in the Western sense. At the end of the first millennium, there are traces of guilds in Nippur, which at that time was a large city. Not only were there traces of guilds, but also of other kinds of association. There were Hindus, Aramaens, and people from Anatolia, with a head man for each group. How they functioned, what their relationship to the government was, I don't know. In the early cities, the craftsmen's associations were within the temple organization. How they developed later is not clear."

Referring to the problem of relations between city and country, Professor Oppenheim remarked that "for heuristic reasons, rather than because it is strictly correct, I have stressed the contrast between city people and people living outside the city. In Mesopotamia there was in fact a tripartite division. We have to differentiate city people—those people actually living within the walls and having a special status—from peasants—people living in small towns and hamlets along canals. The peasant and city-dweller were in about the same relation as the Muslim villager and the Muslim city-dweller. They might very well have had certain cultural and social notions in common. In this way they were different from a third category—the nomads or better, pastoralists.

PART II

The Traditional Muslim City

Introduction to Part II

Immense reaches of time separate the ancient Babylonian city from the Middle Eastern cities of Islamic times. By the advent of Islam in the seventh century A.D., the whole of the Middle Eastern region from the Nile to the Jaxartes had become citified. Virtually everywhere, Middle Eastern peoples relived the Mesopotamian experience in which the complex civilization of towns superseded neolithic village life. The common experience of city life, however, did not eliminate the legacy of local traditions, social ideals, and religious beliefs and practices which formed unique local cultures.

On the other hand, successive empires of increasing magnitude and ever more profound influence—the Babylonian, the Assyrian, the Achaemenid, the Parthian, the Roman, and the Sassanian empires—had forged and diffused the elements of a common Middle Eastern society and culture. Knowledge of a common language, common literary traditions, religious conceptions, law, commercial and administrative practices, and common views about the nature of imperial authority were the stuff of a common Middle Eastern civilization, especially in cities, and especially among the city elite. Soldiers, governors, administrators, merchants, and priests transcended particular localities to live in an empire-wide cosmopolitan society. Between ancient and Islamic times, the monetization and commercialization of economic life, the emergence of rational bureaucratic organization, and the spread of universal religions also transformed city life, creating new classes, new institutions, and new values, despite the legacy of local traditions and local cultures. A city could be as parochial a place as an isolated village, or a sophisticated center of world culture.

Widespread Middle Eastern cultural and religious movements, as well as the rise and fall of empires, had a decisive effect on city development. The era of Hellenistic influences, beginning in the fourth century B.C., left as its most profound legacy to the history of city life the concept of the city as *polis*. The idea of the city as a union of families constituting a single all-embracing, self-governing community, controlling its affairs by the assemblies of citizens or of designated elders, was a Greek concept. From the late fourth century, in a great wave of urbanization, the *polis* form of city spread as new Greek colonies were founded and their example influenced the evolution of older towns. If the *polis* form itself was not adopted, elements of local privilege and autonomy were encouraged throughout the region.

Yet, however widespread the Greek influence, ancient institutions and societies continued to evolve along lines laid down in Middle Eastern rather than classical antiquity. The weight of the classical municipal heritage, and how long it persisted, is still a matter of contention. For a long time it was believed that the Arab conquests and the advent of Islam, like the barbarian invasions of Europe, ruptured the continuity of antique civilization and debased classical institutions or replaced them altogether. However, as early as the fourth century A.D. both the Roman and the Sassanian empires had radically altered the municipal institutions themselves. Increasingly centralized and bureaucratic administrations reduced local privileges and local autonomy. Bureaucratic centralization progressively eliminated classical civic and social institutions, replacing the independent city corporation with centrally designated authorities, while the triumph of the higher religions, Christianity and Zoroastrianism, shifted the bases of local social life from civic and secular to religious and ecclesiastical identifications. Long before the Arab-Islamic conquests, the *polis* had ceased to exist. The Arab conquests did not destroy the ancient city, but confirmed and continued changes begun in late antiquity.

The impact of the Arab conquests and the Arab empires on Middle Eastern cities has yet to be adequately appraised. On the one hand, the Arabs created a number of new and very important cities to settle the conquering tribal peoples, and the Arab empires subsequently lent great impetus to urbanization and the extension of city settlement in many parts of the Middle East. Expanding agricultural and commercial activities, as well as political and administrative needs, favored city growth. On the other hand, it is well to remember that most Middle Eastern cities were ancient foundations preserving, if not the legacy of the *polis*, the qualities of late Hellenistic Roman and Persian town life. Traditional culture was slow to change in the early Islamic epoch. Ancient traditions were preserved, sometimes in defiance of Islam, sometimes by the subtle insinuation of ancient mores, values, and practices into the fabric of Islam itself.

Broadly speaking, the first three centuries of the Arab era, to the collapse of the 'Abbāsid empire in the middle of the tenth century, seem to have formed one major period in the evolution of city life. The breakup of the empire, and the substantial completion of the conversion of Middle Eastern peoples to Islam, created a new era of city experience in the eleventh and twelfth centuries. In his illuminating paper, Professor Grabar suggests a more nuanced, though tentative framework for the periodization of city development in the Muslim age.

Focusing on the one characteristically Muslim institution, the mosque, Professor Grabar first defines the mosque as it was understood by Muslim writers and then examines the phases of the evolution of mosques as revealed in literature but, more particularly, as discovered by archeological investigation. Professor Grabar suggests four stages: a primitive or early phase before 650 A.D.; the classical mosque of the Umayyad era 660–750 A.D. marked by the close identification of the mosque and Arab imperial authority; a third period, from the ninth century, of the separation of the mosque from political control; and finally a fourth phase, which Professor Grabar dates as beginning in the twelfth century, in which radical changes in the form and distribution of mosques and the multiplication and variegation of religious structures signal far-reaching changes in Muslim community and city life.

Though the task of correlating these phases with other historical experiences and with the evolution of other Middle Eastern city institutions in the early Islamic period remains to be done, Professor Grabar's article suggests an historical framework which has implications beyond the history of the specific institution with which he deals. The history of the mosque echoes the development of the Islamic community, from its early unity and the close association of religious and secular political affairs, to the emergence of the Muslim empires with regional and even world interests, and to the subsequent alienation of Muslim social and religious from imperial political concerns. The breakup of the 'Abbāsid empire and the nomadic invasions in the following centuries wrought great changes in the fabric of Muslim social and community life. The intensification of local loyalties and the increasing vigor and independence of local religious organizations in the eleventh and twelfth centuries were reflected in the special architectural situation of the later period. Professor Grabar's study of one crucial institution points the way to a history and temporal typology of Muslim cities.

Concentrating on Professor Grabar's fourth period, which he defines as including the epoch from the eleventh to the fifteenth centuries, Professor Lapidus attempts, from the study of literary rather than archeological materials, to define some of the social characteristics of Muslim cities in this epoch. No longer self-governing polities, Muslim cities, like ancient ones, were ruled by kings and emperors and their governors, garrisons, and tax collectors. Nonetheless, Muslim cities enjoyed an organized social life at the level of family and local quarters and in larger religious groups. Religious communities were a basic form of Muslim social organization, but they were not exclusively urban communities. Religious communities united both urban and rural people into comon societies.

In what ways did the religious communities of Muslim cities resemble the temple communities of ancient cities? To what extent were the elders and notables of Muslim cities equivalent to the leading families of the ancient ones? To what degree did the relationships between the city communities and the "palace" or ruler follow the same parameters? Obviously the Islamic evidence shows less of city corporations. Moreover, no privileges were derived from city dwelling. But these questions posed by the juxtaposition of the various studies require further thought. Professor Adams considers some of the important continuities and discontinuities in his concluding remarks.

Turning from social to geographical conditions, Professor Lapidus takes up the questions posed by Professor Oppenheim about relations between city and countryside. Not denying evident differences and antagonisms, Professor Lapidus attempts to show a substantial degree of relationship. Implicity recalling Professor Oppenheim's remarks about the "steady, osmotic interaction" between city people and country people, he lays stress on migration between city and country, the interest of city people in the agriculture of the surrounding countryside, and certain occupational, social, and functional similarities. He suggests that in many cases the functioning settlement of late Islamic times was not the city as such, but larger districts or regions including both towns and villages.

Turning from generalizations based upon Muslim geographical and historical texts, Professor Goitein comes to terms with the study of one city in particular, the great metropolis and emporium of Fusṭāṭ, or old Cairo. Studying neither archeological nor formal literary treaties, but basing his exposition instead on the documents of the Cairo Geniza, a storehouse for Jewish documents which preserved a treasure of private correspondence, contracts, bills of sale, and other personal records, Professor Goitein gives us a concrete picture of the actualities of Cairo life. He discusses the attitude of Jews, as reflected in marriage contracts, to life outside of the capital and notes their strong antagonism to life in the *Rīf*, the provinces. He gives us a concrete account of the physical form of Fusṭāṭ and its economic life, and discusses the zoning of the city between industrial and residential, rich and poor areas, the ownership and occupation of housing, and the prevalence of ruins. Business partnerships are shown to be a crucial determining factor in the physiognomy of the city.

Professor Goitein goes on to describe some aspects of the political and social life of old Cairo, in particular the organization of government, police power and taxation. Among his interesting remarks is the conclusion that guilds did not exist in Fusṭāṭ, as they did not, we may add, exist in Egypt until the Ottoman conquest. In conclusion Professor Goitein points out that despite the absence of a specific concept of citizenship, Cairo's residents felt deeply attached to their cities and their homes.

In these three papers, based on different types of evidence and different methods of study, we have the basis of a history of the traditional Muslim city as seen through its archeological remains, a typology of city geographical and social organization, and a concrete study in depth of the actualities of life in one particular city. The conferees were agreed that whatever the virtues and limitations of any particular approach, only a pluralism of methods could deepen our knowledge of so complex a phenomenon.

OLEG GRABAR

The Architecture of the Middle Eastern City from Past to Present: The Case of the Mosque

The significance of mosques in the definition of an Islamic city-pattern is acknowledged by all general theories of the Islamic city and is also obvious to anyone who has visited a contemporary Middle Eastern city. Not only are mosques a feature common to all Muslim cities, but they also have had a continuous existence and exert a profound influence upon the cities.

Moreover, since the mosque can be studied in the context of many different urban settings, it provides an advantage over the other two methodologies commonly used in studies of Middle Eastern cities. The first one postulates the existence of an "Islamic" city pattern. Apparently justified by an urban ideal going back to the Prophet's *ḥadīth*, by the existence for several centuries of a large unified empire which, directly or indirectly, sponsored cities from North Africa to Central Asia, and by a number of literary sources, this view has led over the past thirty years to a number of more or less systematic and more or less extensive statements about the "Muslim city." Louis Massignon, George Marçais, Edmond Pauty, Leopoldo Torres Balbas, and Gustave von Grunebaum have in various articles been the most lucid exponents of the notion of a pan-Islamic urban order, in which a large number of local peculiarities are overshadowed by a community of purposes and of habits of life—at least during the classical centuries of Islamic civilization in the Middle East.

The second approach is the precise study of local conditions. This type of information may involve monographs on individual modern cities, such as Clerget's study of Cairo, or investigations of the development of a given city over the centuries, such as Sauvaget's *Alep*, or the identification of a key moment in the history of a city, such as Le Tourneau's study of Fez under the Merinids or Mantran's Istanbul in the seventeenth century. This approach is also used in the study of institutions or characteristics which affect several cities; for instance, the various studies devoted by Claude Cahen to social organisms or by R. Brunschvig to fairs.[1] The information provided by these documents is peculiar in several ways. It is fragmentary since it involves only a limited number of towns, and it will never be complete. Even the most optimistic scholar cannot envisage the possibility of monographs on all cities of the Middle East or on all pertinent and comparable institutions. Furthermore, this information is interdisciplinary to an almost frightening extent: archeology, epigraphy, traditional philology, art history, and virtually all branches of the social sciences are involved in its formulation. Finally, it is intellectually unsettling because its validity—the degree to which a precise bit of knowledge about a specific city at a specific time can be used for anything but that city at that particular time—is often difficult to determine.

In contrast to these two methods, an examination of the mosque helps to bridge the gap between generalizations about the "Middle Eastern city" and specific monographic data. The theme of the mosque or, more generally, of the building which reflects the religious needs of Islamic culture has already been broached in many studies. Both archeological and literary information on mosques is plentiful, and there is a large scholarly literature on the subject. However, it is not yet possible to develop a full and thorough statement of what the mosque has meant to the Muslim city. Therefore this paper will present a sketch of what seem to be the main features of the historical development of the religious building in Islam. Through the study of this particular feature as it changes over time, some meaningful conclusions about the city should emerge. But I should like at the very outset to emphasize how much is still uncertain and hypothetical about the method I will use as well as about many precise details.

What is a mosque? Let us turn first to textual evidence. The most pertinent passages from the Koran seem to me to be the following ones:

> II, 144: And now verily We shall make thee turn (in prayer) toward a *qibla* which is dear to thee. So turn thy face toward the

[1] A complete bibliographical apparatus is unnecessary. Two works can lead the reader to most of the significant studies or sources. One is G. E. von Grunebaum's masterful essay, republished in *Islam, Essays in the Nature and Growth of a Cultural Tradition*, (London, 1955), pp. 141–58. The other one is Ira M. Lapidus' recent *Muslim Cities in the Later Middle Ages*, (Cambridge, 1967), with a particularly complete bibliography, especially on pp. 239–41.

> *masjid al-ḥarām* and ye (O Muslims), wheresoever ye may be, turn your faces (when ye pray) toward it.
>
> XVII, 1: Glorified be He who carried His servant by night from the *masjid al-ḥarām* to the *masjid al-aqṣā.*
>
> IX, 17–18: It is not for the idolaters to tend God's sanctuaries [*masājid*], bearing witness against themselves of disbelief . . . . He only shall tend God's sanctuaries who believeth in God and the Last Day and observeth proper worship and payeth the poor-due and feareth none save God.
>
> IX, 107–108: And as for those who chose a place of worship [*masjid*] out of opposition and disbelief, and in order to cause dissent among the believers, and as an outpost for those who warred against God and His messenger aforetime, they will surely swear: We purposed naught save good. God beareth witness that they verily were liars. Never stand (to pray) there. A place of worship [*masjid*] which was founded upon duty from the first day is more worthy than thou shouldst stand to pray therein, wherein are men who love to purify themselves.
>
> LXXII, 17: Verily sanctuaries [*masājid*] are but for God.
>
> XXII, 40: Sanction is given for fighting to those who have been expelled from their homes unjustly because they said: our Lord is God. For had God not repelled some people by means of others, churches [*ṣawāmi*], synagogues [*bīʿ*], oratories [*ṣalawāt*], and *masājid* would have been destroyed.[2]

From these passages no clear conception of a specificially Muslim sanctuary or temple emerges. The word which ties all of them together is the word *masjid*, but it is not necessarily a building for the new faith (except possibly in the very obscure last passage quoted); it is merely a place which is generally defined as belonging to God. The matter is of particular interest because it contrasts with ritual obliga-

2 Each of these quotations poses a different problem of exergesis, especially those from *sūras* IX and XXII which are clearly related to precise incidents (see commentaries or summaries of discussions in R. Blachère, *Le Coran*, 3 vols., (Paris, 1947–1951). Additional information may be gathered from such books as Ibn Isḥāq's *The Life of Muhammad*, tr. A. Guillaume, (London, 1955), which suggest a slightly more complex as well as more specific context for the mosque in the Prophet's time than I propose here. See also M. Gaudefroy-Demombynes, *Mahomet*, (Paris, 1957), pp. 108, 120, 202, 522, etc. I still feel, however, that even if it will have to be modified in details, my interpretation of the Koranic evidence is justified, among other reasons, by the later history of the mosque.

tions which are spelled out in far greater detail. Furthermore, while there is no clear Muslim holy building, the Meccan sanctuary is recognized as the central holy place of the faith, with the mysterious *masjid al-aqṣā* as a less clear second sanctuary.[3] Finally, the Koran has no statement which would define the physical character of a *masjid* or which would attribute to it any sort of architectural or symbolic characteristic.

The early *ḥadīth* and whatever is known of the practices of the early Muslim community before and a few decades after the Prophet's death provide a few additional data about early Islamic sanctuaries, but, as is well known, these data are very difficult to situate properly in time. Preliminary investigations of this data seem to point to the existence of several partially contradictory trends in the early Muslim community. One was the notion that prayer is an individual act and thus, to paraphrase a celebrated tradition, a *masjid* exists wherever one prays. An even more important result of this direct relationship between man and God, as it expresses itself in the act of prayer, is the lack in Islam of any clergy or intermediary between Creator and creature. Both of these notions tend to make a building with complex ritual requirements unnecessary.

On the other hand, a third early feature of Islam has an opposite result. It involves the complex notion of the community of the faithful (already apparent in the celebrated passage from the Koran LXII, 9–11, calling to prayer on Fridays), with its concomitant features such as the *khuṭba* and its symbol the *minbar*, the choice of Friday as the main day of gathering, or the appointment of specific hours for formal prayer and the growth of a ceremony of the call to prayer. These are the features which permitted the slow transformation of the Prophet's house in Medina into a sanctuary, a phenomenon for which there is no evidence in the Prophet's own time.[4] It is also out of the notion of the community of the faithful that there arises the most characteristic, if not the only characteristic, requirement of the early Muslim sanctuary: a large enough space for the whole body of the faithful who find themselves in any one place.

[3] O. Grabar, "The Umayyad Dome of the Rock," *Ars Orientalis*, III (1959), with further references.

[4] There is as yet no definitive study on these early problems of the cult. See J. Sauvaget, *La mosquée omeyyade de Médine*, (Paris, 1947), especially pp. 134 ff; C. H. Becker, "Die Kanzel im Kultus des alten Islams," and other studies in his *Islamstudien*, I (Leipzig, 1924); S. D. Goitein, various articles gathered in *Studies in Islamic History and Institutions*, (Leiden, 1966), especially chapters IV and V.

One last element must be added to this equation of early Islamic needs for a mosque even though I know of no clear textual evidence for it. It is the existence of churches and synagogues identified with other systems of faith. As the conquest took place, Christian sanctuaries with their highly developed architecture and complex symbolism sometimes served as positive, but more often as negative models for the Muslims. While the sting of rejection by organized Jewish and Christian communities led the Muslims to adopt cult practices which differentiated them from Jews and Christians (the most obvious example is that of the *qibla*), numerous individual conversions and cultural osmosis created a constant influx of internal suggestions for the adoption of Jewish and possibly even Christian habits and practices.

The need for a space large enough to contain the community of the faithful, the principle of the individual act of prayer, the presence of Jewish and Christian traditions, and, except for Mecca, the lack of any concrete notion of a holy sanctuary seem to be the only features which can be proved or assumed to have existed in early Islamic times. The definition of these features derives totally from literary sources since no archeological information is available for the period preceding the conquest. There is perhaps some danger in drawing too many conclusions about this period, for the sources tell us more about what the culture wanted to do than what it actually did. Altogether then, it appears impossible to say precisely what a mosque was in early Islamic times even on the assumption that there was a clear conception of a mosque.

Let us turn now to the late Middle Ages and to the most celebrated theoretical formulation of the Muslim world, Ibn Khaldūn's *Muqaddima*. Its first characteristic for our purposes is that the chapter dealing with mosques identifies only three monuments as being mosques: the sanctuaries of Mecca, Medina and Jerusalem.[5] The chapter closes with the following remarkable statement:

> The ancient nations had mosques which they venerated in what they thought to be a spirit of religious devotion. There were the fire temples of the Persians and the temples of the Greeks and the houses of the Arabs in the Ḥejaz, which the Prophet ordered destroyed on his raids. Al-Mas'ūdī mentioned some of them. We have no occasion whatever to mention them. They are not sanctioned by a religious law. They have nothing to do with

[5] Ibn Khaldūn, *The Muqaddima*, tr. F. Rosenthal, vol. II (New York, 1958), pp. 249 ff.

> religion. No attention is paid to them or to their history. In connection with them, the information contained in historical works is enough. Whoever wants to have historical information (about them) should consult (the historical works).[6]

All sanctuaries, past and present, Muslim or not, are simply dismissed as fakes in the eyes of God.

This is not to say that Ibn Khaldūn does not have anything to say about mosques. But the chapter in which he discussed them is not the one which concerns itself with places of worship but the one dealing with the *imāmate*.[7]

> The leadership of prayer is the highest of (all these functions) and higher than royal authority as such, which, like (prayer) falls under the caliphate. This is attested by the (circumstance) that the men around Muḥammad deduced from the fact that Abū Bakr had been appointed (Muḥammad's) representative as prayer leader, the fact that he had also been appointed his representative in political leadership. They said: 'The Messenger of God found him acceptable for our religion. So, why should we not accept him for our worldly affairs?' If prayer did not rank higher than political leadership, the analogical reasoning would not have been sound. If this is established, it should be known that city mosques are of two kinds, great spacious ones which are prepared for holiday prayers, and other, minor ones which are restricted to one section of the population or one quarter of the city and which are not for the general attended prayers. Care of the great mosques rests with the caliph or with those authorities, wazirs, or judges, to whom he delegates it. A prayer leader for each mosque is appointed for the five daily prayers, the Friday service, the two festivals, the eclipses of (the sun and the moon), and the prayer for rain. This (arrangement) is obligatory only in the sense that it is preferable and better. It also serves the purpose of preventing the subjects from usurping one of the duties of the Caliphs connected with the supervision of the general (public) interests. The (arrangement) is considered necessary by those who consider the Friday service necessary, and who, therefore, consider it necessary to have a prayer leader appointed.
> Administration of the mosques that are restricted to one section of the population or to one quarter of the city rests with those

6 Ibn Khaldūn, II, 266.
7 Ibn Khaldūn, I, 449–50.

who live nearby. These mosques do not require the supervision of a caliph or ruler.

There appears in these paragraphs a rather curious *ex post facto* recognition of the existence of sanctuaries and an attempt simply to record the legal position of a phenomenon which is not explained and perhaps not fully sanctioned. Ibn Khaldūn refers to al-Māwardī, who also deals with mosques in one of his chapters on the *imāmate* and who is somewhat more explicit. He recognizes two kinds of sanctuaries (*masājid*): official ones (of which he names three types: *masājid* proper, *jawāmi'*, and *mashāhid*) which are controlled by the caliph or his representatives; and private ones (*'āmmiyya*) which are the responsibility of whatever person or group built them. al-Māwardī's concern is primarily a legalistic one, that of defining properly the validity of the fundamental Muslim act of prayer.[8]

From Ibn Khaldūn and al-Māwardī we acquire new information in addition to what was provided by literary evidence dealing with early Islam. First, the mosque appears to be legally defined primarily as a place for prayer, and a hierarchical value is given to each place of prayer according to its relationship to the institution of the *imāmate*. Second, a distinction seems assumed—although neither theoretician has been willing to discuss it in any detail—between divinely ordained sanctuaries (of which there are only three), the only true *masājid*, and man-created places for worship. Within the latter, al-Māwardī distinguishes three types but does not define them. This point leads us to a third conclusion: whatever the theoretical constructs of Muslim scholars of the Middle Ages, there appears in them a certain uneasiness about institutions, practices, and buildings which had developed and yet did not seem to fit into the "system." This point has been made more than once with respect to political institutions. It is interesting to note that it applies also to monuments of religious architecture.

We should also consider another type of literary source—descriptions of cities and of their monuments. These accounts give us some indication of the actual, physical reality of religious institutions in the Muslim world, without our relying exclusively on the chance preservation of specific monuments. Even these accounts have to be used with some care whenever one attempts to generalize about the Muslim world. It is only from about the twelfth century on that they ap-

[8] al-Māwardī, *Les statuts gouvernementaux*, tr. E. Fagnan, (Algiers, 1915), pp. 209 ff.

pear in sufficiently large numbers to permit generalization; earlier accounts are either valid for one city only[9] or are too brief to be really useful, as are the classical geographers. For archeology or the history of art, however, the later sources are invaluable for they are the only sources which permit us to appraise the historical and documentary value of surviving monuments. The contrast between the picture provided by city descriptions and theoretical statements is quite striking. Al-Maqrīzī in the fifteenth century lists the following religious or primarily religious buildings in Cairo: eighty-eight *jāmi'ṣ*, seventy-four *madrasas*, nineteen *masjids*, twenty-one *khānqās*, twelve *ribāṭs*, twenty-five *zāwiyas*, three *mashhads*; thirty-three *masjids* and one *jāmi'* were found in the suburb of Qarāfa.[10] Ibn 'Asākir describing Damascus in the twelfth century lists 241 *masjids*, twelve *madrasas*, and one *ribāṭ*.[11] Similar data from Aleppo, Jerusalem, Baghdad, Isfahan, or Samarkand would illustrate the point that, whereas Muslim theoreticians saw the religious institution as consisting of three divine sanctuaries with a hierarchy of man-devised spaces (both known as *masjids*) officially recognized only for the specific purposes of private or communal prayer, the actual development of religious institutions in Islam was far more complex. It included a differentiation of functions, illustrated by the growth of a terminology for religious buildings which does not appear in Ibn Khaldūn or al-Māwardī, as well as a transformation over the centuries of the physical character of the city. This is evidenced by the fact that the small number of functionally defined buildings in early Islam was followed by a tremendous multiplication of structures with religious purposes.

At this point, I would like to turn to the archeological evidence in order to consider how an architecture inspired by the needs of the faith developed. Preliminary investigations of various aspects of this subject have led me to suggest four major periods in the growth of religious architecture in Islam.

It is clear that all early Islamic cities had what we may call in today's parlance a Muslim "civic center." In the newly created cities, which are better documented, these centers developed in two stages which are quite close to each other in time and yet quite different in

9 The most celebrated one is the Khaṭīb's *Ta'rīkh Baghdād*, of which there will soon appear a new translation with important commentaries by Professor J. Lassner.

10 al-Maqrīzī, *al-Mawā'iẓ wal-I'tibār bi-Dhikr al-Khiṭaṭ wal-Athār*, especially vol. II of the undated Cairo edition.

11 Ibn Asākir, *Ta'rīkh Madina Dimashq, ed. S. al-Munajjid*, (Damascus, 1951).

significance.[12] The first stage (Basra in 635, Kufa in 639, Fusṭāṭ in 641–642) consisted of the creation of a sort of forum, open from all sides and directions (partial exception in the case of the Egyptian city) somewhere in the center of the city. This forum was usually called a *masjid* and in one instance a *muṣalla*.[13] It served all the functions which affected the *jamā'a*, the community, from prayer to military recruitment to collection of taxes. To this Muslim center corresponded a group of tribal centers, also called *masājid*. In later times, as is suggested in a text transmitted by al-Maqrīzī,[14] these early creations were interpreted as part of a coherent plan in the Caliph 'Umar's mind. 'Umar was said to have forbidden parallel development of Muslim and tribal institutions called by the same name in Syria. 'Umar is also supposed to have decreed that only one *masjid* may be founded and that individual tribes must be prevented from building their own. Whether or not this was consciously planned by the second caliph, it is indeed true that in the old cities conquered by the Muslims (information is available for Damascus, Hama and Jerusalem) a single Muslim entity was created. Usually it was begun by taking over some disused or little used open space near the center of the city and, after minor repairs, this space functioned in the same fashion as the Muslim *masjid* in the new cities.

The main characteristic, then, of this first stage was the creation of a space which served exclusively Muslim purposes and which, in cities that were entirely Muslim, existed on two separate levels of exclusivity. The word *masjid* is always associated with these spaces, but it does not yet possess any formal structure nor does it have any precise function other than that of excluding non-Muslims.

A second stage occurred between 650 and 750.[15] To my knowledge, twenty-seven *masjids* from this period are archeologically definable. This figure includes modifications to earlier buildings, but excludes buildings known through texts only (this unfortunately means all Iranian mosques). If we bar from consideration such local topograph-

12 The main evidence has been gathered by K. A. C. Creswell, *Early Muslim Architecture*, vol. I, (Oxford, 1932), chapters 1 and 2. Some of these schemes, especially as they have been put together so brilliantly by L. Caetani, *Annali dell 'Islam*, (Rome, 1905–1918) appear at times too neat and may already have been medieval simplifications. But even if this is so in part, the pattern suggested makes sense.

13 al-Baladhūrī, *Futūḥ al-Buldān*, ed. M. de Goeje, (Leiden, 1866), p. 350.

14 al-Maqrīzī, II, p. 246.

15 In this division my position departs somewhat from that of K. A. C. Creswell, *Early Muslim Architecture*, and J. Sauvaget, *La mosquée omeyyade*. These scholars separate the changes which occurred between 650 and 700 A.D. from those which followed al-Walīd's rule. A detailed discussion of the reasons for my position cannot be made here.

ical features as may have affected individual changes during this century, the following points seem to characterize this period. First, each Muslim center continued to have *masjids*, but the tendency was to recognize only one for each place: local tribal *masjids* still existed, but their importance dwindled. In the case of Basra, for instance, physical changes were made in the *masjid* of the community because Ziyād Ibn Abīhī, the governor, was afraid of the undue importance taken by smaller *masjids*. No other term than *masjid* appears to have been used, although instances occur of the word *muṣalla*, but these instances (especially in Medina) seem to refer to an institution *extra muros* which still demands investigation. Second, the *masjid* was transformed from a space into a building. This is a crucial development of this century and is all the more remarkable because all twenty-seven mosques were related in form. They were all hypostyle constructions with columns or piers as the main units of construction and bays framed by two or four columns or piers as the module which allowed an almost infinite growth of the building in any direction. In this respect the early mosque was a remarkably modern building which could be expanded or contracted according to the needs of the community.[16] All mosques had a certain relationship between open and closed covered spaces. The problems posed by this relationship pertain primarily to the history of art, except on one point, which is the apparent tendency to consider the covered parts as the *bayt al-ṣalāt*, i.e. place of prayer, and the rest of the building as an overflow area for prayer. All these buildings were enclosed by walls and did not have an exterior façade. Their orderly form appeared only from the inside where the balance between open and covered spaces served, among other things, to indicate the direction of *qibla*. Their only outward symbol was the minaret, a feature which appeared early in mosques built in old cities with predominantly non-Muslim populations and only later in primarily Muslim ones. The minaret was only one of several new features found in all or most mosques; others were the *miḥrāb*, the axial nave, the *maqṣūra*, the decoration, a small dome in the center, and so on. All these features can be explained as due to various secular needs,[17] but all of them tended during this period to acquire a religious meaning or, to be more precise, a cultic meaning by becoming involved in the ceremony of prayer. Yet more importantly, these mosques, like Con-

[16] On this point see O. Grabar, "La mosquée de Damas et les origines de la mosquée," *Synthronon*, Paris, 1968.

[17] Most of them have been studied by J. Sauvaget, *La mosquée omeyyade*. See also H. Stern, "Les origines de la mosquée," *Syria*, XXVIII (1951), and O. Grabar, "Islamic Art and Byzantium," *Dumbarton Oaks Papers*, XVIII (1964), 74.

stantinian basilicas, were almost all willful creations of princes and of governors. They were closely tied to palaces and to the *dār al-imāra,* and were rarely spontaneous creations reflecting the immediate spiritual or ritual needs of the populations. If it is too strong to refer to them as imperial mosques, their consistent formal typology and their use as models for later times certainly permit us to call them "classical" mosques.

During the second half of the eighth century, as well as during the ninth and tenth centuries, the "classical" mosque type dominates the whole Islamic world. It is the time of the great masterpieces of Cordova, Samarra, Kairouan, Cairo, and Baghdad. The very same type seems to have existed in Iran, although our information is too scant to permit certainty on this point. With the growth of huge metropolises such as Baghdad, Samarra, and Cairo, some cities acquired several *masjids* with equal legal status. The increase of large sanctuaries is usually explained as being due to the increase of population, but other factors are involved as well. Imperial glory was a factor in the construction of the mosque of Ibn Ṭūlūn, the Azhar, or the Samarra mosques, while the development of local social identifications in Baghdad's population also contributed to the growth of several other mosques.[18] However, sheer size was the main factor and each mosque probably served as a center around which the life of various city sections was organized.

Thus, partly as a consequence of these divisions within large urban units, a less immediate relationship prevailed than had previously existed between the mosque and the caliph or his governors. Palace and mosque were no longer necessarily adjacent. Officials appeared less often in the sanctuaries. In Fāṭimid Cairo, the caliph's visit to the four mosques of the agglomeration became an organized and carefully regulated ceremony instead of a common occurrence.[19]

At the same time the considerable development of such features in the mosque as the *miḥrāb* area or the so-called T pattern can be explained as the results of purely religious, almost spiritual values attributed to the mosque.[20]

[18] For Baghdad see note 3, above. On other cities the most accessible sources of information are K. A. C. Creswell, *Early Muslim Architecture,* vol. II, and *Muslim Architecture in Egypt,* vol. I, (Oxford, 1952).

[19] See the description of the inauguration of al-Ḥākim's mosque, al-Maqrīzī, II, pp. 280 ff.

[20] The theme deserves further study; in the meantime see E. Pauty, "L'évolution du dispositif en T dans les mosquées à portiques," *Bulletin d'Études Orientales,* II (1932).

While the formal typology of the *masjid* during these centuries remained more or less as it had been in the first century of Islam, and while no obvious major changes seem to have occurred in the function of *masjids,* the mosques in the very large cities became partly dissociated from the secular authorities and developed as specifically religious symbols. In lesser towns the situation varied considerably. In Cordova the palace was still adjacent to the mosque. In Bukhara or Merv, on the other hand, the governor's palace seems to have been independent of the mosque. (There is some uncertainty as to the position of the mosque in the large *maydāns* which began to appear.)[21] The situation in small towns or villages is hardly known. Altogether, the exact characteristics of this third period are not easy to establish and, pending the discovery of new material, I would prefer to define it as a period marked by refinement in the internal arrangement of the classical mosque with significant novelties demonstrable only in the very large cities.

A fourth period can be fixed on archeological grounds as belonging to the twelfth century, but for a number of reasons which still require study, it probably began somewhat earlier. One of the most important aspects of this period were the changes in the appearance of the whole Muslim world, changes which were not the same throughout but varied with local traditions.

Let us examine the archeological evidence. The major monument of Islamic architecture of the late eleventh and early twelfth centuries is the Great Mosque of Isfahan.[22] Its very complicated history and its extraordinary aesthetic merits need not concern us here. It is important to know, however, that this is the first known example of a series which remained typically Iranian until today. An aerial view of its location within the city shows a position clearly akin to that of the mosque of Damascus and of the Ibn Ṭūlūn mosque in Cairo. It is fully integrated within the city and obviously occupies a large space. It has no clear outside façade and can be entered from several places. The internal arrangement, however, is changed. The twelfth-century mosque replaced an earlier hypostyle mosque and reflected, therefore, a conscious formal change. Instead of the large space of the hypostyle hall with its endless possibilities of movement and growth, there is an interior courtyard (and not merely the open part of a single area),

21 For the Central Asian examples useful indications and summaries are found in G. A. Pugachenkova and L. Rempel, *Istoriia Iskusstva Uzbekistana,* (Moscow, 1965).

22 There is still no usable monograph on this building. Best summaries are by A. Godard in *Athar-é Iran,* I and II (1936–1937).

with an interior façade and a division of the covered parts into four separate areas through the creation of large *eyvāns* on each side of the courtyard. The earlier internal unity of spatial arrangement is gone, and enlargements become impossible except through the addition of separate buildings attached to the original *masjid.*

In Cairo during the same period, in addition to the four large mosques previously discussed, several small mosques acquired spectacular features such as a street façade that awkwardly fitted into the pre-established pattern of the city.[23] Mausoleums appeared in the city and especially outside its walls and some of these had small oratories attached to them.[24] In Damascus, Aleppo, Mosul, and Baghdad we can not only follow the same developments but also we can witness the rather sudden appearance of a hitherto unknown type of building, the *madrasa. Ribāṭs* and monasteries of various types also appeared in large numbers within the cities, although these forms were previously found mostly in frontier areas. A major terminological change followed. The term *masjid* tended to refer only to the smaller sanctuaries, while *masjid al-jāmi'* and later simply *jāmi'* referred to the older or larger ones. The latter word eventually took over as the only word for mosque. The exact moment when the linguistic shift occurred is not very certain. The earliest formal occurrence known to me is in the text of an inscription (but unfortunately not a building inscription) dated 956 A.D. and copied by al-Maqrīzī.[25] Among writers I have consulted, Ibn Ḥawqal (late tenth century) seems to be the first to use the term systematically. In this case, however, I am uncertain whether the word is a colloquialism or whether it corresponds to an official terminology.

In any event, these novelties imply two major changes in the structure of the urban system. On the one hand, varying architectural forms evolved to serve Muslim piety. Communal prayer in a large mosque may have continued; but, parallel to the mosque, mausoleums for holy men and women also appeared, as well as private oratories which were identified with smaller social units (family, quarter, profession). Muslim versions of monastic orders which separated some individuals from the total community, along with a new system of teaching and training in the faith which was separated from that of the traditional mosque, also called for new structures. In the large

23 Creswell, *Muslim Architecture in Egypt*, pp. 239 ff.

24 Creswell, *passim*, and O. Grabar, "The Earliest Islamic Commemorative Buildings," *Ars Orientalis*, VI (1967).

25 al-Maqrīzī, II, p. 255.

mosques themselves—such as that at Isfahan—the breakup of the original single unity of the building can be explained as a result of divided allegiances within the city; the community no longer prayed together but formed smaller groups for prayer.

This change in the nature of the community of the faithful, indicated by an analysis of the monuments, suggests many different hypotheses. I should like to single out several in particular. First, the *morcellement* of spiritual allegiances, when related to the grandeur of the many new buildings, shows a widening of the social base of architectural patronage. More people acquired more means to build more numerous and more varied types of pious buildings than ever before. This extension of patronage and of taste can be confirmed by evidence from other arts.[26] But these new constructions also adapted themselves to the existing pattern of the city. They no longer transformed the city by becoming its obvious centers but fitted themselves wherever space was available. Indeed the earlier city had often imposed small and sometimes awkward shapes upon them. Alternately, new sanctuaries moved outside city walls and were one of the contributing factors to the growth of suburbs, as has been shown by Sauvaget's study of Damascus.[27]

A second change implied by these architectural novelties is perhaps even more important for an understanding of the structure of the city. The cultic and spiritual life of the city was no longer tied to one or to a few large places but to a vast number of buildings. In this respect the city of the twelfth century appears to have consisted of a series of parallel and probably partly competing poles of spiritual allegiance and religious behavior. While to my knowledge there never occurred a parish-like organization in Islam, archeological evidence suggests that the allegiance of the individual Muslim was parochial, though it is not clear whether the parochialism was related to quarters or whether certain city-wide organizations took precedence over topographical proximity. To interpret the evidence we need further textual investigation. However, two other archeological phenomena can be added to our dossier. One is the burgeoning of minarets, particularly prevalent in Cairo and Isfahan. Minarets began to be constructed in the twelfth century. In certain places in Iran, minarets

[26] See some preliminary remarks on this subject in O. Grabar, "The Illustrated *Maqamat* of the Thirteenth Century: The Bourgeoisie and the Arts," to be published by A. Hourani and S. M. Stern in the proceedings of the 1965 seminar in Oxford on the Islamic city.

[27] J. Sauvaget, "Esquisses d'une histoire de la ville de Damas," *Revue des Études Islamiques*, VIII (1934), 461.

still remain even though their mosques have disappeared. Minarets were hardly necessary in such quantities for the specific aim of calling to prayer; rather, like the spires of churches, they were symbols of the presence, not so much of individual religious institutions, as of the people who built them or for whom they were built. Like the façades of mausoleums or of other sanctuaries, minarets became a form of conspicuous consumption and publicity for the buildings with which they were found. Thus they contributed to the creation of monumental avenues, like the *shāri' bayn al-qaṣrayn* in Cairo, where a whole series of superb buildings exemplified the same needs and functions. The other phenomenon is the accentuation of a trend we have noted in the previous period. Secular buildings become completely separated from the sanctuaries, and, even when certain holy places were found in the citadel itself, as in Aleppo or Cairo, these places no longer played a significant part in the spiritual life of the city.

The sketch I have proposed for the fourth period in the development of the religious building in Islam still requires a few additional remarks. First, it seems that this phenomenon is not valid for the Muslim West. Second, the phenomenon lasted in the Arab world through the Mamlūk period and in Anatolia until the beginning of Ottoman power. In Iran, with the advent of the Mongols, large imperial complexes took over, as they would do later under the Ottomans. These complexes usually included most of the functions of the buildings of the twelfth and thirteenth centuries, but little is known as to whether they were actually used in the same way or whether they were merely expressions of imperial glory. Third, after the fall of the Mamlūks, many of the institutions for which all these buildings were built fell into disuse; in Iran mosques and *madrasas* disappeared leaving only minarets and mausoleums.

Still, the usefulness of the monuments of this period was not exhausted. Here perhaps the documentation I have presented for the Middle Ages may serve those who try to understand the contemporary city. For not all the buildings were destroyed, and, as new institutions developed, especially in the nineteenth century, these old buildings were restored and employed anew. It is no accident that old religious *madrasas* in Damascus are used as girls' schools, libraries, and academies or that the Suleymaniya, in Istanbul, houses the main photographic laboratory for the library collections of the city. Other such buildings, such as Baybars' mosque in Cairo, are public gardens, for their large space is perfectly suited to the city planner's concern for air and greenery. How restful and clean are the fountains in the courtyards of the many medieval buildings, in which today as in the past one escapes from the noise and the dirt of the city! In these ways the various medieval religious developments which I have outlined have provided Middle Eastern cities—almost none of which are new cities—with a monumental frame or grid, in a manner comparable to the ways in which the Roman city created the grid of the medieval Mediterranean city. In part, the needs for open spaces in the modern city were already answered in the Middle Ages, although not for the same reasons of health; and the mosque, small or large, with its court is clearly far more "contemporary" in its function than the closed church or cathedral of continental Europe. The expression of the complexity of the urban structure in numerous architectural monuments, beginning around the twelfth century, and the decadence of the structure after the fifteenth century, which did not necessarily entail a destruction of the monuments, has provided many a Middle Eastern city with spaces and often buildings which can be reused by contemporary organs of government, society, and culture.

However fragmentary and incomplete they may be, the information and the hypotheses which have been presented lead to a number of conclusions. First, a study of an Islamic architecture of religious inspiration indicates the existence of an evolution, of which we have defined four stages. For an understanding of the city—and especially of the pre-modern city—this evolution has several implications. One is that there was more than one *type* or *model* for the traditional city and that these types are definable in chronological succession. Another implication is that the pre-modern city acquired its essential characteristics around the twelfth century; the earlier, more unified city was superseded by a city with a multiplicity of spiritual allegiances, whose exact mode of operation still demands detailed studies.[28] This city in turn also decayed, but it did provide the contemporary town with architectural nuclei which can be reused for contemporary purposes.

A second conclusion derives from the fact that, except for the earliest periods, most of the documentation presented here has been archeological. This evidence does not appear to coincide with theoretical statements made by classical Muslim writers. Does this mean that the behavior and attitudes suggested by the monuments were so obvious that writers did not record them? Or does it mean that the

28 See now the book by I. M. Lapidus, cited in note 1.

reality of religious life was radically different from official statements about the faith? In any event, it would appear that for an understanding of the growth and development of the medieval world, sources derived from the fields of material culture may often be more authentic and more valuable than traditional literary ones.

Finally, these remarks are also intended to suggest problems and subjects for further study. There is a methodological problem concerning the exact validity of a scheme based on only one aspect of urban life for the study of the whole city. There is a linguistic problem concerning the history of the terminology used for religious architecture and institutions.[29] And there is the problem of equalizing our knowledge of the various areas. This is particularly true when one considers that we know so much more about the Arab world than about Iran or Turkey.

But these problems will not be solved by one person; they require a long range systematic effort by teams drawn from various fields of concentration and with different linguistic competences. Without such an effort we will always end up caught between excellent studies on points of detail, on the one hand, and brilliant but untested hypotheses, on the other.

DISCUSSION

The members of the conference were particularly interested in some of the ramifications of Professor Grabar's remarks about the development of cities in his fourth period, the twelfth to the fifteenth centuries. Professor von Grunebaum considered the cultural context for this period and noted that both Professor Grabar and Professor Lapidus point out, for the period after 1200, "the domination of the city by clerical circles, by the law schools, and by people who created a particular atmosphere in education and knowledge. This is known, but as far as I can see, insufficiently exploited in one context with which we all are intrigued; and that is the question of why and when the intellectual impetus in Muslim civilization, or rather in Arabic civilization, dried out. It seems to me that you have given one additional clue to the many clues which we have had so far. When you look back from that period to the ninth and tenth centuries, which were the high-point of the Muslim intellectual movement, there was

[29] It should be pointed out that none of the Arabic terms mentioned in this paper have ever been studied in an historical fashion.

an *adab* (I don't mean *adab* as good manners, or *adab* as the ability to quote poems, but *adab* as a cultural ideal, as a formal ideal, as an ideal of behavior) which connected and tied together the courtly circles, the ruling circles, and the intellectual circles. By the twelfth, thirteenth, and fourteenth centuries, the domination of the city's intellectual atmosphere by the concerns of the clergy marks the end of any intellectual model which would bind together the ruler and the ruled. The fact that the majority of these rulers were foreigners, and were only gradually and imperfectly assimilated, points this out."

Professor Lapidus remarked on the sociological aspects of the changes in building patterns, the development of new structural types, and the great multiplication of religious structures. He agreed with Professor Grabar that this implies a fragmentation of community life, but thought that "some distinctions have to be introduced. The multiplication of religious institutions did not necessarily imply an intensification of the very small scale community ties of quarters or fraternities. While the single mosques of the classical period implied a unified society most of the new structures seem to have represented instead the growth of resources, prestige, and power in the large religious communities which stood outside of small quarters and groups.

"Socially speaking, these buildings also represented the patronage of the various military regimes and their interest in supporting and furthering, for a complex of motives, the activities of the religious community. I think Professor von Grunebaum's remarks about the breakdown of a common culture between the military and intellectual and clerical circles is very germane in this respect. Genuine social ties, those of common cultivation, were replaced by formal external ties of alliance, support, and patronage. I would suggest that although we do have a fragmentation of community life, the multiplication of institutions ultimately represented the enhancement of activity in the larger religious community as supported by political regimes."

Professor Goitein raised the question of what concrete meaning a mosque had: "How closely were people attached to a particular mosque? Were they attached to a mosque as such or to the *imām* or prayer leader? I believe, for example, that many indications of a mosque belonging to a certain group of people, such as the mosque of the coppersmiths and so on, do not really imply a mosque for a corporation, but rather a mosque in a particular quarter or market." Professor Fakhry responded: "Muslims have no preference whatever; all mosques are the house of God, and if anyone has a certain attachment to a particular mosque, it is a matter of convenience. You are also quite right that the name indicates nothing at all and may be used simply because the quarter has that name even if there are no coppersmiths living there. People may go to a particular mosque but it has nothing to do with their profession. They go because of the good man they find leading prayers or preaching there."

Professor Fernea, on the basis of contemporary experiences, saw other possible explanations for the multiplication of mosques. He observed that the "completion of a monumental structure such as a mosque requires both a cultural tradition for models and conceptions, and the organization of political, economic and social forces to realize this model. It seems to me that these two elements, the cultural tradition and the sociological background for the organization of the work, cannot be taken to be the same through such long periods of time and over such a wide geographic area. I think of two contemporary examples. I was doing some work among the Bedouins in northern Saudi Arabia a couple of years ago, and I happened to visit the town of Sākākā. As I was walking through the town, I noticed numerous *masjids*. I counted some two hundred and was surprised to find afterward that in a town of about 20,000 to 30,000 people, there were approximately 400 to 500 *masjids*. It turned out that these numerous *masjids* were by and large not used. They were really a device which enabled the king, through his ministry, to provide an income for as many religous people as he could. It was a way of providing a kind of welfare pension for his subjects. I found this apparently true also in al-Qurīyāt and in Jidda.

"The other example that comes to mind is from Nubia. Beginning with the 1900's, Nubia experienced what is perhaps the most remarkable architectural renaissance of any rural area in the Middle East. From a collection of very small houses built close to the river, perhaps one room to a family, Nubian villages changed into conglomerations of houses with double courtyards, elaborate facades and fine decorations. Elaborate colonades going down to the river front appeared in the villages. Along with this renaissance of domestic architecture and the increasing size and ratio of numbers of rooms to people, a large number of mosques were built, some of them with double minarets or with domes.

"There were certain economic circumstances involved. One was compensations that the government provided for loss of land when the first dam at Aswan was built. Another was enforced idleness, since the rising of the reservoir cut the agricultural season to half of what it had been. In addition, villages moved up the mountainside where land was not as valuable for agriculture. All of these things help explain the economic circumstances of this development.

"There was a further consideration: in the southern half of Nubia, the mosques, but not the houses, were much more modest in size and complexity than the mosques in the northern part. We later found out that the villages of the northern part of Nubia, where a different dialect from that of the southern part is spoken, are not only residential units but also tribal units, whereas the people of the southern part of Nubia are not tribally organized; their villages are purely residential units and there is a great deal of fractionation of lineage groups. Thus it seems that the people who built the elaborate mosques in the northern part of Nubia had residential interests, tribal connections, and tribal leadership to organize the kind of communal life necessary to build this architecture, whereas it was lacking in the other half of Nubia where all the effort went purely into the domestic work." Professor Fernea concluded that the sociological and cultural considerations which explain such phenomenon as the proliferation of religious institutions in the twelfth century may be complex indeed.

Professor Grabar was thus led to consider, apart from his original sociological explanation, the possibility that these buildings were a form of conspicious consumption, investments on the part of princes, *Mamlūks*, merchants, and others, intended to impress people or tie down funds within certain families or groups. "We may have an architecture which corresponds not to the structure of the society, but to the interests of certain people or groups within it."

Coming to more specific points, Professor von Grunebaum reconsidered the question of whether there were mosques in the time of Muḥammad, in the earliest Muslim community. "In *The Life of the Prophet* by Ibn Hishām the episode of the destruction of the so-called mosque of the opposition is recounted. The story is that the apostle went out until he stopped in a town an hour's daylight journey from Medina. The owners of the mosque of the opposition came to the apostle and asked him to come and pray with them. The prophet, however, did not go there but sent people to the mosque of those evil men to destroy it. They went and took a palmbrush and lighted it, and then two of them ran into the mosque and burned it. Thus, it seems that we do have a mosque in the prophet's time."

Professor Grabar noted that palm trunks with palm leaves were used for protection against the sun, "but I have the impression that a mosque as a Muslim building, different from others, a place where God is worshipped, was not clearly differentiated in very early times. There is still no technical Muslim meaning of mosque."

IRA M. LAPIDUS

Muslim Cities and Islamic Societies

INTRODUCTION

Broadly speaking, the study of Muslim cities has been founded upon the assumption that Muslim cities are self-contained entities which comprise a distinct society and culture, radically different from and opposed to that of the peasantry and the countryside. Though they do not resemble the Greek or Roman *polis* or the medieval European commune, it is thought that some aspects of administrative and religious organization, or perhaps a particular quality of life, nonetheless make Muslim cities total or comprehensive societies. Strong currents of ideology and emotion support this view. For Muslims, cities often possess a special sanctity and are regarded as the sole places in which a full and truly Muslim life may be lived. Muslim urbanites are deeply attached to their residences, intensely despise peasants and peasant life, and insist on the superior virtues of the cities—sentiments in accord with European and American preferences.

Yet, as I will attempt to show in this paper, the belief in the unity of city societies, and the conviction that city and country are radically opposed, are exaggerated and misleading. The nature of the Muslim city society and its relationships to the world around bear re-examination. I should like to consider cities not as isolated artifacts, but in terms of their relationships to the larger social, geographic, and religious environments in which they are embedded. In the first part of the paper, I shall examine the structure of medieval Muslim social groupings and the relationships between social bodies and city spaces; in the subsequent parts, I shall consider the geographical and religious aspects of city organization and city relationships to rural areas. A reconsideration of social structures, geographical forms, and religious organizations suggests new approaches for understanding the nature of cities in the Muslim world. In dealing with these problems, I concentrate on the medieval age of Islam in the ancient core regions of the Muslim world—on the period from the late tenth to the fifteenth centuries in the region between the Nile and the Jaxartes. Despite the many differences and changes in the regions of the former 'Abbāsid empire, the period between the disintegration of that empire and the consolidation of the Ottoman and Safavid empires has, for reasons too complex to consider here, a unity of its own. It is perilous to make generalizations for such an immense region and period in which the cities discussed differed from one another and changed internally in numerous important ways. Nonetheless, certain features of geography and social structure seem to have been held in common, and differences in detail do not vitiate all efforts at generalization. Though historians are disposed to stress the immediate, the concrete, and the unique in historical experiences, models or "ideal types" which define the whole as well as assess the significance of differences in detail are equally important.

I attempt here to present such a model of Muslim social organization in the post 'Abbāsid period. The object of this typology is not to describe any particular situation in all its manifold reality, but to propose a number of categories which make the historical sources more meaningful. As touchstones of thought, these categories reveal the degree to which any individual case resembles or differs from the model. The model also facilitates systematic comparisons between different situations. Naturally, the model is not an a priori construction; rather it is abstracted from a detailed study of the particular circumstances which it closely describes, and is tested and refined by whatever additional information is available. However, even within these limits the remarks made in this paper are hypotheses based upon a limited survey of the sources. Accordingly, they are advanced only as tentative efforts to interpret the available data and to suggest further possibilities for inquiry. By no means are they conclusive formulations.[1]

[1] Before proceeding to detailed notes, I wish to acknowledge the stimulation I have found in the literature and from colleagues in this field. G. E. von Grunebaum's "The Nature of the Islamic City," reprinted in *Islam: Essays in the Nature and Growth of a Civilization*, (London, 1961) is the main point of departure for any study of Muslim cities. Various works by C. Cahen including *Mouvements Populaires et Autonomisme Urbain*, (Leiden, 1961) are essential for defining some of the main problems in the field, and for path-breaking discoveries about Muslim social life. C. E. Bosworth's *The Ghaznavids*, (Edinburgh, 1963) is rich in historical and sociological appreciations of Nishapur and Iranian cities in the eleventh century. Jean Aubin's observations on the structure of the Iranian *pays*, to be published in the proceedings of a conference on the Islamic city held at Oxford in June, 1965, have stimulated me to try to think outside

MUSLIM COMMUNITY STRUCTURES

The forms of Muslim community life pose complex questions about which we have little knowledge, and studies of Muslim social structure are few and incomplete. For this reason, I would like to begin with a brief review of my findings about the social structures in Aleppo, Damascus and other cities of the Mamlūk period. Other evidence suggests that the social characteristics of these cities were not atypical. In the present state of our knowledge, the situations in Aleppo and Damascus may serve not as description, but as models for the analyses of other cities in the Muslim world during the Middle Ages. Though each place differed in detail, certain formal organizational qualities were held in common.

In Aleppo and Damascus the basic units of society were quarters, which were social solidarities as well as geographical entities. Small groups of people who believed themselves bound together by the most fundamental ties—family, clientage, common village origin, ethnic or sectarian religious identity, perhaps in some cases fortified by common occupation—lived in these neighborhoods. Moreover, quarters were also administrative units headed by a *shaykh* who was appointed by the city governor to assist in taxation, maintain order, enforce police ordinances, and represent the quarter on city-wide political or ceremonial occasions. Quarters were village-like communities within the urban whole. Indeed some quarters were suburban districts, composed of people of recent village or bedouin origin.

These sharply divided city populations enjoyed relatively few institutions which cut across quarter boundaries to bind them together. Guilds or other merchant, artisan, and professional groupings were extremely weak. They were usually created to meet the regulatory and fiscal needs of the Mamlūk state, and rarely reflected the autonomous interests of their members. Somewhat more effective in bridging quarter divisions were various fraternal associations—Sufi brotherhoods, youth clubs, and criminal gangs. Such fraternities, however, were not an exclusively urban phenomenon. For example, *zu'ar* gangs were also found in the surrounding villages and were sometimes allied with the *zu'ar* of the city proper. Though socially and po-

the usual urban-rural dichotomy. Charles Tilley, now of the University of Toronto, helped me crystallize the notion of the city as a center of functions rather than of communities. Not least am I indebted to the participants of this conference for many valuable suggestions and criticisms.

litically important in both towns and countrysides, such solidarities were marginal to the rest of society and fell short of providing a basis for the integration of the populations as a whole into a single community.

What larger communities existed in Mamlūk Aleppo and Damascus were created by the *'ulamā'*, the learned religious elite. Their schools of law were socially as well as religiously central. In religious terms, a Muslim school of law expounded the body of legal and moral teachings (based on the Koran, the sayings of the prophet Muḥammad, and the consensus of Muslim jurists) which constitutes the ultimate expression of Muslim beliefs and code of proper behavior. Socially speaking, a law school consisted of the group of scholars and teachers who elaborated and preserved the law and the witnesses and judges who supervised its implementation in the community at large. The schools were built around the more or less formally organized *'ulamā'* study groups—circles of scholars, students, admirers, and patrons, along with the clientele of notaries, orderlies, and clerks in the service of the *qāḍīs*. From these core memberships, the schools reached out to include the populace at large. All Muslims were members of one or another of the schools in that they looked to the *'ulamā'* for authoritative guidance on how to live a good Muslim life, for judicial relief, and for comfort and leadership in times of trouble. Insofar as the *Sharī'a*, the Muslim holy law, defined social obligations, the *'ulamā'* administered the social and economic as well as the purely religious aspects of Muslim town life. More concretely, family ties, the close association of the *'ulamā'* with officials, merchants, and artisans, who were recruited from all quarters and classes of the population, bound the people to the schools and created communities beyond parochial quarters—communities which shared a common law, common norms in family, commercial, and religious life, a common judicial authority, and common facilities such as mosques, schools, and charities.

The communities created by the schools were communities for ritual, familial, commercial, educational, and legal purposes. Though more inclusive than parochial quarters and fraternities, they were not, however, states or governing bodies. They possessed no power to tax. Their funds were derived from gifts and endowments. They possessed no organized military force, had little formal administration, and held no jurisdiction over territory. Nor were the schools of law city or city-wide communities. In Aleppo and Damascus, not one, but several of the four equally orthodox schools were represented. In no sense was a school of law a substitute urban polity.

Both the smaller *'aṣabiyyāt* and the law schools of Aleppo and Damascus were encadred in the Mamlūk state organization. A slave military caste, monopolizing military power and controlling fiscal administration, dominated the territories in which the smaller communities were located. The Mamlūk empire provided military protection, manipulated the local economy, gave political and social support to the *'ulamā'* elites, and helped resolve conflicts between the various community bodies. A large territorial empire complemented the social and political capacities of the smaller communities.[2]

The fundamental elements of Mamlūk period social organization—the quarter, the fraternity, the religious community, and the state—seem to have prevailed throughout the Muslim world, from Egypt to Central Asia, from the eleventh to the fifteenth centuries. Almost universally, Muslim cities contained socially homogeneous quarters. Such quarters were found in cities created by a coalescence of villagers, by the settlement of different tribes, or by the founding of new ethnic or governmental districts. Quarters based on the clienteles of important political or religious leaders, religious sects, Muslim and non-Muslim ethnic minorities, and specialized crafts, were also found in cities throughout the Muslim world. Even such tiny minorities as foreign merchants might have their own quarter, in the form of a *fundūq* or caravansary set aside for their residence and business. However cosmopolitan, the great cities were no exception to this rule. The populations of ninth and tenth century Baghdad, for example, lived in separate sectors. The caliphs assigned each ethnic regiment in their forces to its own district, while migrants from Basra and Iranian and central Asian cities had their own markets or quarters as well. Religious groups such as the Ḥanbalīs, Shī'is, and of course Christians, were also identified with distinct parts of the city. Though less coherently or less exclusively organized elements may have been present in city populations, neighborhood communities seem everywhere to have been the keystone of Muslim urban life.[3]

2 For fuller discussion see I. M. Lapidus, *Muslim Cities in the Later Middle Ages*, (Cambridge, Mass., 1967).

3 Baghdad: G. Le Strange, *Baghdad during the Abbasid Caliphate*, (Oxford, 1924); G. Salmon, *L'Introduction topographique à l'histoire de Baghdād*, (Paris, 1904); M. Streck, *Die Alte Landschaft Babylonien nach den Arabischen Geographen*, (Leiden, 1900), pp. 175–239; Ya'qūbī, *Kitāb al-Buldān*, (Leiden, 1906), pp. 233–54. Sāmarrā: P. Schwarz, *Die 'Abbāsiden-Residenz Sāmarrā*, (Leipzig,

Beginning in the eleventh century, the integration of such divided populations into more highly ordered communities seems to have taken the form of religious communities. These communities resembled the law schools of Mamlūk Aleppo and Damascus, and may be regarded as forerunners and even prototypes of the later schools. In both chronicles and geographies, the populations of towns are identified by their religious affiliations, though ethnic identity is occasionally mentioned. For example, Rayy, Nishapur, and Bukhara were basically Ḥanafī cities while Merv, Kazvin, Shiraz, Ardebil, and Kazerun were basically Shāfi'ī.[4]

Further we know that by the middle of the eleventh century, the Muslim *'ulamā'*—once essentially a religious elite—had emerged as the social and political elite of cities throughout the Muslim world. The process by which this occurred is obscure. However, it can be said that the old order of society was swept away by the disintegration of the 'Abbāsid empire in the tenth century, and by the establishment of new Turkish regimes, as well as regimes dominated by slave soldiers, in the eleventh century. The military, administrative, and landowning classes of the earlier empire were replaced by new elites who took control of armies and administrations throughout the region of the former 'Abbāsid empire.

Only one element in the past complex of elites—the Islamic religious leaders, the *'ulamā'*—survived the political traumas of the tenth and eleventh centuries. Independent of the empire and protected by the widening acceptance of the religion whose official interpreters

1909); Streck, pp. 47–171; Ya'qūbī, pp. 255–68. Kufa: L. Massignon, "Explication du Plan de Kūfa," *Mélanges Maspero*, III (1940), 337–360. Basra: L. Massignon, "Explication du Plan de Basra," *Westostliche Abhandlung R. Tschudi*, pp. 154–74; C. Pellat, *Le milieu basrien et la formation de Ǧāhiz*, (Paris, 1953). Cairo: Nāsir-i Khusrau, *Sefer Nameh*, (Paris, 1881), pp. 144–45; P. Ravaisse, "Essai sur l'histoire et sur la topographie du Caire," *Mémoires par les Membres de la Mission Archéologique Française au Caire*, (Paris, 1887), 409–80; A. R. Guest, "The Foundation of Fustat and the Khittahs of that Town," *Journal of the Royal Asiatic Society*, 1907, pp. 49–83. Nishapur and other Iranian cities: C. E. Bosworth, *The Ghaznavids*, (Edinburgh, 1963). Syrian cities: C. Cahen, *Mouvements Populaires et Autonomisme Urbain dans l'Asie Musulmane du Moyen Age*, (Leiden, 1961). See also Gibb and Bowen, *Islamic Society and the West*, vol. I, part I, (Oxford, 1950), 276.

The geographers indicate quarters for Jerusalem, Bukhara, Isfahan, Samarkand, Herat, Nasaf, Balkh, Merv, Gurgan, Shiraz, Rayy, Wāsiṭ, Hamadan, and many smaller cities.

4 Ardebil: Mustawfī, *Nuzhat al-Qūlub*, (London, 1919), p. 84. Kazerun: Mustawfī, p. 125. Merv: Nāsir-i Khusrau, p. 275; Yāqūt, *Mu'jam al-Buldān*, VIII, (Cairo, 1906–1907), 35. Nishapur: Bosworth, *passim*. Kazvin: Barbier de Maynard, "Description historique de la ville de Kazvīn," *Journal Asiatique*, X (1857), 264–66, 287; Mustawfī, p. 64. Rayy: Yāqūt, IV, 356.

they were, the *'ulamā'* were able to assume some of the functions of the older administrative and landowning classes. They became a political and social as well as a religious elite. In many cities—first in eastern Iran, Fars, and Azerbaijan, and later in Syria, Mesopotamia, and Egypt—leading *'ulamā'* families merged with landowning, bureaucratic, and merchant families. Such families provided city leadership for generations. Some controlled the judicial and educational administration. Others, such as the Mīkālīs of Nishapur and the Burhān of Bukhara formed dynasties of *ra'īses*—chiefs over administration, police, justice, and taxation. Their power had its source in popular religious followings and in the control of offices and landed incomes. Also, by the eleventh century, the *'ulamā'* had partial control of *waqfs*, which were permanent endowments of incomes for religious and charitable purposes. These endowments had become the economic foundation of the *'ulamā'*, strengthening their position as a professional, judicial, scholarly, administrative and religious elite.[5]

To what extent the Iranian *'ulamā'* elites of the eleventh century formed schools of law like those of Syria and Egypt in the thirteenth, fourteenth and fifteenth centuries remains unclear. *'Ulamā'* leadership, and popular identification with the *'ulamā'*—constituted schools seems certain, though the nature of popular support is problematical. In Baghdad and other places, the Ḥanbalī school seems to have had an organized mass following, whereas the Ḥanafī and Shāfi'ī of Nishapur seem to have enjoyed no more than diffuse popular sympathy and *ad hoc* support. How much these apparent differences, and the differences between these and the later Mamlūk period schools are due to our lack of knowledge, how much to an earlier and incomplete evolution, and how much to basic structural differences remain problems for further investigation.

[5] Save in Bosworth's *The Ghaznavids* and C. Cahen's *Mouvements Populaires* and in my studies of the Mamlūk period, the subject of the *'ulamā's* social and political leadership and the role of the city *ra'īses* has scarcely been broached. R. Frye has important observations on this subject in *Bukhara, The Medieval Achievement*, (Norman, Oklahoma, 1965). Further scattered instances of *'ulamā'* leadership in communal affairs may be found in the geographers but a full study of the subject requires utilization of chronicles and biographical dictionaries. In addition to the materials mentioned above, other scattered materials are also available. *Qāḍīs of Shiraz*: G. Le Strange, "Description of the Province of Fars," *Journal of the Royal Asiatic Society*, 1912, pp. 14–15, 316–17. Other *ra'īses*: Yāqūt, I, 216–17; III, 157; al-Muqaddasī, *Aḥsan al-Taqāsīm fi Ma'rifat al-Aqālīm*, (Leiden, 1906), p. 313; V. Minorsky and C. Cahen, "Le recueil transcaucasien," *Journal Asiatique*, CCXXXVII (1949), 97–105; V. Minorsky, *A History of Sharvān and Darband*, (Cambridge, 1958), pp. 117–18, 124–27.

Another illuminating difference between our model Syrian cities and other cities should be noted. Outside of Mamlūk Syria, orthodox Muslim schools were not the only organized religious communities. In many towns, during the eleventh to fifteenth centuries, Muslim theological sects such as the Mu'tazila, the Karāmiyya, together with various kinds of Shī'is and Ismā'īlis, were socially and politically active as well as religiously represented.[6] It seems probable, though our evidence is more imprecise than in the case of the orthodox schools, that these religious groups were organized communities comparable to the Sunni schools.

Thus far we have spoken of the law schools as urban bodies, but the importance of Muslim religious groups impels us to consider the aspects of Muslim religious organizations which transcend individual town or city situations. We know already that the Muslim law schools were not really town bodies in a territorial sense. Within cities a single school might come to dominate a particular place, and for all intents and purposes create a kind of communal unity despite the presence of minority schools. But in many, if not most cases, the schools represented factions, often bitterly hostile factions, which shared a city space. For example, Isfahan, Samarkand, and Damascus were divided between Ḥanafīs and Shāfi'īs, and Damascus had a strong Ḥanbalī minority as well. Ḥanafīs, Shāfi'īs, and Karāmiyya were all important in eleventh century Nishapur.[7]

Moreover, looking beyond city space, the schools of law do not appear to have been exclusively urban bodies. Law school jurisdictions extended to rural areas outside of cities. Reciprocally, the populace of rural areas identified with and looked to the town-centered schools for social and judicial leadership. Often a majority, if not the whole population of an oasis or of a district formed by a market town and

[6] Shī'i communities—Rayy: Yāqūt, IV, 356. Kazvin: Mustawfī, p. 64. Basra: al-Muqaddasī, pp. 126–30. Nishapur: Bosworth, pp. 194–200. Shiraz: Mustawfī, p. 113. Qum: al Idrīsī, *Géographie d'Édrisi*, A. Jaubert, tr., II, (Paris, 1836), 167; al-Iṣṭakhrī, *al-Masālik wal-Mamālik*, (Cairo, 1961), p. 119; G. Le Strange, *The Lands of the Eastern Caliphate*, (Cambridge, 1905), p. 209; Ibn Ḥawqal, *Kitāb Ṣūrat al-Arḍ*, II, (Leiden, 1938–39), 361, 370; Mustawfī, p. 71. Nihāwand: Mustawfī, p. 76.

For Ismā'īlis in Syria, see B. Lewis, "The Ismā'īlis and the Assassins" in Kenneth Setton, ed., *The History of the Crusades*, vol. I. For Mu'tazila in the following provinces—Hamadan: Mustawfī, p. 75. Fars: Ibn Ḥawqal, II, 292; al-Muqaddasī, pp. 439, 441; P. Schwarz, *Iran im Mittelalter*, (Leipzig, 1896–1936), IV, 381. Kerman: Schwarz, III, 232. Khuzistan: Ibn Ḥawqal, II, 255; al-Muqaddasī, p. 415; Schwarz, *Iran*, IV, 334, 412.

[7] Isfahan: Yāqūt, I, 273. Samarkand: Mustawfī, p. 238. Damascus: Lapidus, p. 112. Nishapur: Bosworth, chapter V.

its surrounding villages, a *nāḥiyya* or *rustāq*, shared membership in a school of law. Fortified by the other social and political interests, this religious identification made a real, rather than a nominal, community. Towns were often the center of communities which included both rural and urban dwellers.[8]

In other cases, probably more common, cities and their hinterlands were divided among several schools or sects, the rural areas echoing the city situation. In the twelfth century, Shāfi'īs, Ḥanafīs, and Shī'is made up the population of Rayy, with corresponding religious groups being represented in the surrounding countrysides. Shī'is, Khārijīs, and other sectarian minorities were found in both towns and villages of predominantly Sunni areas such as Fars, Iraq, Mesopotamia and Kerman. Parts of Azerbaijan had mixed Shī'i, Shāfi'ī and Ḥanafī populations. In all such cases, the law school was not strictly a city, but a district or regional body.

The content of these identifications is hard to fathom. No doubt they differed from class to class. Among the *'ulamā'* elites, the interests of the schools probably bound town and village-dwelling counterparts. Among the common people, we cannot say how these ties were conceived. But we do know how strongly they were felt. A dramatic example of the depth of communal bonds comes from the history of twelfth century Rayy. Rayy was harried by communal warfare among its Shāfi'ī, Ḥanafī, and Shī'i populations. The Sunni schools destroyed the Shī'is in the course of these struggles, but then turned against each other. In the ensuing battles, Ḥanafīs from the villages surrounding Rayy came into the city to assist their religious brethren, but the Shāfi'īs gained the upper hand in the town despite the intervention of the Ḥanafī countryside. Similarly, social struggles at Basra and Isfahan united townspeople and village people in the common cause. Social struggles here were not formed on urban-rural but on religious-communal lines.[9] Religious identifications were evidently profound and intense.

Other evidence suggests that religious ties were either based upon

[8] Districts are described as inhabited by a particular religious group without further distinction of town and villages. al-Muqaddasī, pp. 318, 367; Mustawfī, pp. 67, 70–73, 85, 87–88, 94, 126, 148, 150–52, 156; Schwarz, *Iran*, III, 255; Yāqūt, III, 503. Of course, in the many instances in which our sources indicate only "city" populations, they are likely to mean the whole of the attached rural areas as well.

[9] Azerbaijan: Mustawfī, pp. 78–90. Iraq: al-Muqaddasī, pp. 126–27. Fars: al-Muqaddasī, pp. 439, 441. al-Jazīra: al-Muqaddasī, p. 142. Kerman: al-Muqaddasī, p. 469; al-Iṣṭakhrī, p. 99; Ibn Ḥawqal, II, 312. Rayy: Yāqūt, IV, 356. Isfahan: Yāqūt, I, 273. Basra: al-Muqaddasī, pp. 129–30.

or reinforced by other ties between city and country people. Among the elite classes, landowning interests or educational experiences bound village to urban families. City merchants sometimes lived in villages, and rural families often sent their sons to be educated in the cities and to dwell in the *madrasas*, which by the eleventh century not only provided schooling, but also lodging and stipends for these students. Many students remained in the cities, but others returned to their villages to become prayer leaders, notaries, and judges, and to carry on local Muslim life. Among the lower classes, factional ties bound city and village dwellers. In the Damascus region in the Mamlūk period, young men's gangs recruited from intra-urban quarters, suburban quarters, and village districts sometimes joined together to resist the Mamlūk regime. In Syria, tribal and kinship ties also created factional alliances which were comprised of both rural and urban residents, while in eleventh century Iran other *'aṣabiyyāt* took no cognizance of urban or rural residences.[10] All over the Middle East, local market towns or cities—where the *'ulamā'* and the elite of the community were concentrated, where the chief *qāḍīs* resided, where students were trained, where fraternities were centered, where peasants came to market—drew the population of their regions into a single interlocking social body conceived in religious terms. However superior the functions of the towns may have been, Muslim communities were often regional rather than urban bodies.

Until other sources are explored, however, we remain ill-informed about the precise character of urban-rural religious ties. They bound people together in troubled times, but what was the routine quality of the affiliations? Did rural adherents of a law school hold the same cultural, doctrinal, and religious suppositions as did urban members? Were they directly under the jurisdiction of the chiefs of the schools? Did they participate to the same degree as did urban members? To what extent were the law school functions concentrated in towns and to what extent did they exclude villagers? Until such questions are

[10] Townsmen living in rural areas are described above. Landowners, of course, lived in towns (Bosworth, pp. 158, 160–61; Ibn Ḥawqal, I, 30, 138, 141, 211, 215–16; II, 313, 335, 353, 358, 362; al-Muqaddasī, pp. 275–77; Schwarz, *Iran*, I, 34; Ya'qūbī, pp. 246, 274–75, 279, 321–22). Students: Le Tourneau, *Fez in the Age of the Marinides*, (Oklahoma, 1961), pp. 20, 25. Biographies of *'ulamā'* from rural villages are noted throughout Yāqūt. Examples of *'ulamā'* who returned to villages are given above. Examples of *'ulamā'* of village origin but keeping city careers are also numerous. See: Yāqūt, I, 52, 100, 191, 206, 225, 256, 299, 369–70; II, 94–95, 109, 249, 413–14, 432; III, 34, 95, 407, 473; IV, 24, 57.

For *'aṣabiyyāt*, see Lapidus, pp. 87, 157; Professor Jean Aubin, oral communication, and discussions below.

adequately answered, we know only the outward social forms, but not the inner meaning of these religious identifications to the individual Muslims involved.

In stressing the socio-religious bonds between townsmen and villagers, another and apparently contradictory situation has thus far been neglected. In some instances we find townsmen organized into religious or ethnic communities differing from rural religious (or ethnic) groups. In the period following the Arab conquests, Muslim cities were isolated in Christian, Zoroastrian, or pagan countrysides. In the early centuries of Islam before the massive conversion of Middle Eastern peoples, when most of the Muslims were the Arab conquerors and their clients, the towns were identified with an ethnic and religious elite and the countrysides with a population belonging to another society. For example, in North Africa, the cities had long tended to be Arab and the rural areas Berber. In Tabaristan, the cities were Muslim administrative and trading centers amidst local non-Muslim peasant or pastoral peoples. In non-conquest situations, we find orthodox Khurasanian or Transoxanian towns surrounded by heterodox and Shī'i communities or conversely, Shī'i Kashan isolated in a Sunni landscape.[11]

Thus urban-rural division has long been taken to be the typical form of Muslim urban-rural social relations. Many scholars have held that Muslim cities stood apart from, and were intrinsically opposed to, the countrysides which surrounded them. But according to the testimony of the geographers, such urban-rural divisions were evidently exceptions to the rule of religious-communal bonds between town dwellers and the peoples of their hinterlands. The social antagonism between town and countryside is probably best regarded as the result of special circumstances, such as the Arab invasions, or as a special instance—only accidentally urban-rural in form—of a more general kind of social antagonism. Village-city differences resemble conflicts between city quarters, between neighboring villages, and conflicts which united city dwellers and villagers against other parties similarly composed. For example, urban-rural Sunni-Shī'i divisions were no more pronounced than Sunni-Shī'i divisions *within* many

[11] Daylam: Minorsky, *Sharvān*, p. 83. Kashan: Yāqūt, VII, 13; Mustawfī, p. 72. Khurasan: Bosworth, pp. 163–65; R. Frye, tr., *Narshakhī, The History of Bukhara*, (Cambridge, 1954), p. 75. Sāvah: Mustawfī, p. 68. These illustrations are not conclusive for they do not preclude the possibility that rural Shī'is or heretics have urban minority counterparts, and that there are community ties across urban-rural lines even though town majorities may differ from rural majorities.

Iranian and Syrian towns in the eleventh century.[12] Even when communal conflict was simultaneously "urban-rural" conflict, the "urban-rural" can still be viewed as a special variation. Although the conflict may appear to be "city" versus "non-city," only a more general notion of communal organization can adequately account for the forces behind the great variety of other communal antagonisms. We do not yet understand the forces which governed the division of nearby places into hostile communities. But neither spatial considerations nor theories of intrinsic urban-rural antagonism can begin to describe the economic, social, and religious factors involved.

Thus far I have considered the law schools in a city or regional context. In larger dimensions, however, schools of law were super-communities which not only embraced towns and their hinterlands, but extended over the territories of states and empires. Muslim rulers created hierarchies for the *'ulamā'*, appointed the chief *qāḍīs* and professors and the chief Sufi *shaykhs*, and defined their jurisdictions and those of their deputies and subordinates. At both a local and an imperial level, Muslim states defined the spheres within which the legal, educational, and religious activities of the *'ulamā'* schools could be carried out.

In another sense, Muslim schools of law ultimately transcended the states which defined their jurisdictions and existed as world-wide communities wherever Muslims were settled. These communities were united by respect for the most prominent judges and teachers and by the travels of Sufis, students and scholars to and from the major centers of learning and religious activity. Simply put, wherever common law, religious teaching, and traditions were recognized, Muslims were brothers. Not city walls, but natural regions, political circumstances, and cultural identifications delimited the relationships which made effective religious communities.

At various levels Muslim social life was organized in the form of religious communities. However, to avoid confusion it should be clear that in speaking of the schools of law as communities we are speaking of social bodies whose functions differ according to the scale of the territory we have in mind. At the local level the school of law was the organized body in which Muslim community life was

[12] Intra-urban *'aṣabiyyāt* are discussed by Bosworth, Cahen, and Lapidus; see especially al-Muqaddasī, p. 326. For conflicts between cities, see: Schwarz, *Iran,* IV, 356, 364–65, 367; Abū Dulaf, *Travels in Iran,* (Cairo, 1955), p. 41; Yāqūt, I, 53; V, 21, 313; Bosworth, 168–69. Conflicts between villages seem to have been endemic in Syria (Lapidus, pp. 90–91), and are widely attested for modern Egypt.

structured. At the level of empires the ties which made up the schools involved the organizational, financial, and jurisdictional interests of the *'ulamā'* proper. Here the distribution of judgeships, professorships and *waqf* resources was determined. Beyond the empire level, school ties depended on more purely scholarly and religious considerations and on the movement of teachers and students. At this level, not the administrative organization but the common doctrine of the schools took shape. Each level of organization sustained the activities of the school as a whole. Doctrine was no less important to the cohesion of the local communities than local communities were important for servicing the activities of the *'ulamā'*. The law schools, like other communities, were complex organisms which defined themselves at different levels by different functions. Their unity was not derived from formal organization nor from a unified movement, but from the inter-relationships of the *'ulamā'* with each other and with imperial political elites and local populations at the various levels of school activity.

Finally, to return to the comparison of Mamlūk and other medieval social forms, the last organizational element we have to consider are the Muslim empires. However alien in origin and function, imperial regimes played an important role in the formation of Islamic societies. The military regimes which dominated the Middle East from the tenth century on generally kept peace between the various communities, but when states were weak communal violence was rife. The religious communities depended on the military regimes to keep order and equity among them. Although it is too far from our main theme to examine this in detail, we may note that the dominance of military regimes and the corresponding *'ulamā'* stress on obedience reflected the inability of the religious communities to establish political order.

Thus, some city dwellers were identified with the cosmopolitan world of Islam while others were identified with the states or empires which transcended individual localities and coordinated the relations between them. City people were not exclusively attached to their places of residence, but to persons and institutions throughout the larger society.

Thus in the period which followed the collapse of the 'Abbāsid empire and preceded the consolidation of the Ottoman and Safavid empires, there are four kinds of groups or communities: neighborhood bodies; fraternities, including youth gangs and Sufi brotherhoods; religious communities such as the Muslim schools of law and sectarian groups; and, finally, states or empires. None of the bodies was a city community, except insofar as cities were naturally the headquarters of all groups. Some of these groups, such as families, quarters and fraternities were usually small segments of city populations while religious communities and states were much larger than individual cities. As opposed to the ancient *polis* or medieval European commune, there were no geographically defined communities in the Muslim world. Though tentative, this model of social organization suggests that we reconsider the meaning of cities in Islamic societies. If Muslim communities were not exclusively urban communities (even though cities, being large settlements, were naturally crucial foci of social life), how shall we understand the view widely held among scholars of Islam that Muslim cities were opposed to the countrysides? How may we reconcile our conception of social organization with the equally widespread idea that Muslim cities were of crucial religious significance for Muslim societies? Can we hold simultaneously that Muslim social groupings were not urban bodies and that Muslim cities had a distinctive social, geographical, and religious reality? Obviously this is a complex problem for which I have no comprehensive solution. Nonetheless, we might re-examine the geographical and religious aspects of the usual conception of Muslim cities, for upon close examination both these conceptions will have to be modified in ways which suggest new directions for interpreting the meaning of cities in Muslim civilization.

THE GEOGRAPHY OF CITIES AND REGIONS

At one level the fact that Muslim cities had no internal unity has long been evident. Yet this mystique of cities has nonetheless inspired efforts to define a characteristic social quality for Muslim cities. One justification for the persistent feeling that Muslim cities must form a social world of their own stems from the contrast between city and country. The image of the walled town standing in relief against a shapeless countryside has long inspired scholarly quests for that decisive property which unifies Muslim towns. The image, however, is misleading. When examined closely, the geography and ecology of Muslim cities in relation to the surrounding countrysides proves to be exceedingly complex. In fact, in many situations, no *absolute* distinction between urban and rural habitats may be drawn. This may seem contradictory, and heretical to the stereotype of the Muslim city as isolated from and opposed to the countryside. Nonetheless, just as both environments were knit together by elements of social organization, so in certain cases did they interpenetrate or resemble each other in some geographical and ecological aspects. To define precisely what is meant demands fuller consideration of the physical form of Muslim cities and regional geographies.

In the Muslim world of our period, larger settlements such as metropolitan centers, provincial or regional capitals, smaller market towns and even some large villages were not generally distinct entities, but most often composites of lesser units. Settlements of all types, from the largest metropolises to the smallest towns and villages, were clusters of distinct physical and social units.

The great capital cities, Baghdad and Cairo—homes of cosmopolitan populations, creative sources of Muslim culture, centers of imperial administration and international trade—were unique in size but not atypical in form. Cities such as Baghdad and Cairo, with populations estimated at 200,000 or 300,000—vastly larger than any which had existed previously in the Middle East, and several orders of magnitude bigger than their contemporaries—were not single cities but composites of cities. Both developed by the juxtaposition of a succession of palace centers and military encampments, each of which grew into a settlement having the size and characteristics of a separate city. Baghdad, the Madīnāt al-Salām, surrounded by al-Ḥarbiyya on the north and commercial al-Karkh on the south, faced similar districts, such as al-Ruṣāfa across the Tigris, and thus repeated the configuration of Sassanian Madā'in—a city of cities. Cairo, similarly, took shape as a set of adjacent administrative, military, and commercial centers, sometimes separated by open spaces. The most important founding, al-Qāhira in 969 A.D., was neither the first nor the last major extension of metropolitan Cairo. Al-Fusṭāṭ, al-'Askar, and al-Qaṭā'i' preceded it, and subsequently Saladin's citadel and the new districts built around it extended the metropolitan complex into new areas.[13]

Lesser cities were similarly composed. Twin cities or double cities made up of wholly distinct physical entities, often separated by open

[13] See J. Lassner, "Massignon and Baghdad: The Complexities of Growth in an Imperial City," *Journal of the Economic and Social History of the Orient,* IX (1966), 1–27. Lassner regards the presence of a *jāmi'* and a chief *qāḍī* as the defining attributes of a Muslim *madīna* or "city," but this requires qualification, as discussed below.

Yāqūt sums up the observation that any one of Baghdad's quarters resembled a city (Yāqūt, I, 215). For Madā'in see: Ya'qūbī, p. 321; Mustawfī, p. 50. Basra was similarly composed of a number of physically separate quarters resembling towns: V. Minorsky, ed., *Ḥudūd al-'Ālam,* (London, 1937), pp. 138–39; Nāṣir-i Khusrau, p. 236.

From another age, seventeenth century Istanbul shows a similar configuration.

space, were common in the medieval Muslim world. Isfahan and Raqqa, among others, were double cities which slowly grew together across the spaces which separated them. In Egypt and Iraq, many cities and towns were divided into two sectors by canals or rivers, each of which might have separate mosques and bazaars. There were different types of double cities—those composed of fortresses and their suburbs, and those formed whenever suburbs grew in size and facilities to equal the original settlement. Iranian cities generally were composed of several units—a citadel, the city proper, and its suburbs —each surrounded by its own walls.[14]

Divided into three main sectors—Stanbul, Galata, and Uskudar—and again subdivided into palace and governmental complexes, residential and commercial districts, ethnic and class quarters, and surrounded by populous fishing and gardening suburbs, Istanbul reproduced the form of earlier Middle Eastern metropolitan complexes (R. Mantran, *La vie quotidienne à Constantinople, au temps de Soliman le magnifique et de ses successeurs*, [Paris, 1965], pp. 24–42).

14 Double cities formed by adjacent settlement units separated by rivers include the following towns. Adana: Ibn Ḥawqal, I, 183; al-Iṣṭakhrī, p. 47. Ahwāz: al-Muqaddasī, p. 411; Le Strange, *Eastern Caliphate*, p. 234; Schwarz, *Iran*, IV, 321. Hilla: Mustawfī, p. 47; Ibn Ḥawqal, I, 245. Gurgan: al-Idrīsī, II, 180; Ibn Ḥawqal, II, 382; Minorsky, *Ḥudūd*, p. 133. Maṣṣīṣṣa: al-Hamdānī, *Kitāb al-Buldān*, (Leiden, 1885), pp. 112–13; Ibn Ḥawqal, I, 183; al-Iṣṭakhrī, p. 47; *al-Idrīsī*, II, 133; Le Strange, *Palestine Under the Moslems*, (Beirut, 1965), pp. 505–06. Sūq al-Arba'ā': al-Muqaddasī, p. 412; Schwarz, *Iran*. IV, 328; Le Strange, *Eastern Caliphate*, p. 242. Other examples W. W. Barthold, *Turkestan Down to the Mongol Invasion*, (London, 1928), p. 80; Ibn Ḥawqal, II, 420–21; *al-Idrīsī*, I, 369; II, 188–89; al-Muqaddasī, pp. 291, 375–77, 437, 466–67; Schwarz, *Iran*, IV, 246; Le Strange, *Eastern Caliphate*, pp. 242, 317.

Fortresses with suburbs were common in Northern Syria and Upper Mesopotamia on the Byzantine frontier. Examples are Mārdīn and Ḥiṣn Kayfā: Ibn Ḥawqal, I, 224; Yāqūt, VII, 361; Le Strange, *Eastern Caliphate*, pp. 96, 114. Fars: Ibn Ḥawqal, II, 271–72. Also, ports in Syria were divided between the town proper and the fortified anchorage and landing facilities (Nāṣir-i Khusrau, p. 49). Al-Rammāda, outside of Aleppo, is an example of a quarter which became as large as a town, having its own markets and even a separate governor though it touched on the walls of Aleppo (Yāqūt, IV, 282; Le Strange, *Palestine*, p. 519). Al-Ṣāliḥiyya, near Damascus, represented a similar development.

Examples of double cities with a Friday mosque in each sector follow. Isfahan: Ibn Ḥawqal, II, 362, 367; al-Iṣṭakhrī p. 117; al-Idrīsī, II, 167; Le Strange, *Eastern Caliphate*, pp. 203–04; Minorsky, *Ḥudūd*, p. 131; al-Muqaddasī, pp. 388–389; Ya'qūbī, p. 274; Yāqūt, I, 272. Nahrawān: Ibn Ḥawqal, I, 244; Ibn Rustah, *Kitāb al-A'lāq al-Nafīsa*, (Leiden, 1892), p. 163; Le Strange, *Eastern Caliphate*, p. 61; Ibn Serapion, "Description of Mesopotamia and Baghdād," *Journal of the Royal Asiatic Society*, 1895, p. 269. Raqqa: Ibn Ḥawqal, I, 225–26; al-Iṣṭakhrī, p. 53; Le Strange, *Eastern Caliphate*, p. 101; al-Muqaddasī, p. 141. Wāsiṭ: Ibn Ḥawqal, I, 231; Ibn Rustah, pp. 185, 187; al-Iṣṭakhrī, p. 58; al-*Idrīsī*, I, 367; Streck, pp. 318–19. Other examples: Ibn Rustah, p. 186; Le Strange, *Eastern Caliphate*, pp. 237, 242, 318, 484–85; al-Muqaddasī, pp. 196, 273, 410, 412.

Several mosques in the same settlement were rarely found before the eleventh century when *jāmi'* were multiplied in all the major towns.

There are several reasons for the existence of such double cities. Many developed naturally in densely populated areas, especially where waterways or fortified places formed likely foci for adjacent settlements. More dramatically, double cities were created by the Arab conquerors and later Muslim regimes. New suburbs and quarters were frequently developed to settle conquering armies or the ethnic allies of new dynasties in the vicinity of existing city centers. The Arab conquerors of the Middle East not only founded new garrison cities, but also took over parts of established cities or created new suburbs. Later regimes adopted the same policy. Most dramatic was the founding of royal suburbs which housed a ruler's household, administration, and military forces. Baghdad and Cairo were creations of this sort. So too was Shādyākh, near Nishapur. Originally an administrative suburb founded by the Ṭāhirids, Shādyākh was destined after the earthquake of 1154 to replace Nishapur as the main city of the region. Other administrative and military bases in the vicinity of existing city centers were Khurāsānabādh, about a mile from Herat; Wāsiṭ; Rāfiqa next to Raqqa; Fanā Khusraw near Shiraz, which was the palace, military and artisan center of the Buwayhid ruler, 'Aḍud al-Dawla; and numerous places in North Africa—notably Raqqāda near Kairowan and Fāṭimid Manṣūriyya, Muḥammadiyya, and al-Qāhira. Hence we have the double city form.[15]

Moreover, these double multiple cities were also internally divided into separate quarters. In Baghdad and Cairo, each city within the

[15] Shāydākh: Bosworth, p. 161; Mustawfī, p. 147; Ya'qūbī, p. 278; Yāqūt, V, 208–10. Fanā Khusraw: Ibn Ḥawqal, II, 275; Le Strange, "Description of the Province of Fars," pp. 316–17; Le Strange, *Eastern Caliphate*, pp. 249–50; al-Muqaddasī, pp. 429–31; Mustawfī, p. 113. Khurāsānabādh: Ibn Ḥawqal, II, 437; al-Iṣṭakhrī, pp. 149–50; Le Strange, *Eastern Caliphate*, p. 408. Rāfiqa: al-Hamdānī, p. 132; Yāqūt, IV, 208; Le Strange, *Eastern Caliphate*, p. 101; Ibn Serapion, pp. 50–51.

New quarters, built in Rayy by the Caliphs al-Manṣūr and al-Mahdī, in Merv by Abū Muslim and by Ḥusayn b. Ṭāhir (al-Iṣṭakhrī, pp. 147–48) and in Kazvin by the Caliph al-Hādī and his clients, seem also to have been governmental quarters and settlements for administrative personnel.

Other minor examples include a palace suburb of Aleppo built by the amirs of Nūr al-Dīn (Le Strange, *Palestine*, p. 552), and the Rashīdī quarter of Tabrīz built in the fourteenth century (Mustawfī, p. 79). The Banū Mazyad built al-Hilla opposite a previously existing town on the Euphrates at the end of the eleventh and beginning of the twelfth centuries (Ibn Ḥawqal, I, 245; Yāqūt, III, 39). The Khazar capital of Athil was also divided into a government quarter and a popular sector (al-Iṣṭakhrī, p. 130; al-*Idrīsī*, II, 335–36; Ibn Ḥawqal, II, 389). Deeply rooted political and social attitudes and symbols as well as the practical needs of palace, governmental headquarters, and barracks dictated the formation of these suburbs.

metropolitan complex was subdivided into suburbs, quarters, wards, streets, and markets.[16] Lesser cities and towns were also composed of distinct quarters, which were sometimes separated by walls and gates[17] and sometimes merely by the hostile feelings of the community. Such quarters had various origins. Some were vestiges of the circumstances of city foundings. In the cities founded as bedouin encampments, the populace was settled into tribal quarters which, as in the case of Basra, Kufa, Cairo and other cities, retained the characteristics created by the earliest divisions. Similarly, cities founded by the enclosure of a number of villages or by the establishment of a market serving several settlements, might long retain open spaces between the districts and later, though grown into a solid mass, continue to harbor the original settlements as quarters of the city. Muslim Kazvin was formed in this way, as were Qum, Merv, Kazerun, and possibly Bukhara.[18] Most important were the social divisions within city populations which necessitated the creation of separate quarters.

In the formation of such settlements, no absolute distinctions were made between urban and rural elements, or to put it another way, among quarters, suburbs, and adjacent villages. Cities often had an agricultural component. Walled suburbs were often used for gardening and other forms of agriculture, and the outlying villages could be regarded as quarters or suburbs of the city proper. In addition, the fabric of Muslim settlements allowed for gardens and agriculture inside the city proper, either in open spaces or in garden lots attached to town houses. Many cities, especially in Iran, were also surrounded by gardens and fields owned and/or worked by peo-

16 The numerous kinds of subdivisions of Baghdad were called *rabaḍ, maḥalla, murabba'a, sikka, darb, sūq, suwayqa, qaṭī'a,* etc. Similar terms are used for Cairo—*maḥalla, khuṭṭ, darb, sikka, sūq, ḥāra,* etc.

17 The quarters of Isfahan, Nishapur, and Bukhara are said to have been walled, but the quarters of Cairo and Damascus were not permanently fortified before the end of the fifteenth century. Only in time of troubles were gates and barriers erected to protect the inhabitants. See: R. Bulliet, *Nishapur,* unpublished dissertation, Harvard University, 1967; Gibb and Bowen, vol. I, part II, 279; Lapidus, pp. 94–95; Le Strange, *Eastern Caliphate,* pp. 460–62; Nāṣir-i Khusrau, p. 253.

18 Kazerun: Mustawfī, p. 125; Le Strange, *Eastern Caliphate,* p. 266. Qum: Ya'qūbī, pp. 273–74; Le Strange, *Eastern Caliphate,* p. 210; Yāqūt, VII, 159–60. Nishapur: Bulliet. Kazvin: Barbier de Meynard, "Déscription historique de la ville de Kazvīn," pp. 261–62, 273; Le Strange, *Eastern Caliphate,* p. 220; Mustawfī, pp. 62–63. Mustawfī describes the development of Isfahan as due to the coalescence of several villages, but most other accounts discuss only the double cities of Jahūdiyya and Jayy (Le Strange, *Eastern Caliphate,* p. 204; Mustawfī, p. 55). Conversely, Baghdad in its decline reverted to a collection of disconnected settlements separated by fields or wasteland (Ibn Serapion, p. 291; Yāqūt, II, 16, 21, 425; III, 183, 245).

ple who lived in the cities.[19] Thus, the populations of these composite settlements included not only a skilled and sophisticated bourgeoisie of administrators, scholars, merchants, and craftsmen, but also many people who differed little in their attitudes, mores, and manner of life from rural people. Besides agriculturalists, migrants fleeing rural hardship or looking for temporary work came to the cities. Many villagers or nomads settled there permanently, forming quarters or suburbs of their own, while others fell into an unassimilated mass of lumpen-proletarians.[20] Among the middle and upper classes were found nomadic chieftains and people who came from village families to study in the *madrasas* and schools of the capitals. Though the rural upper classes were more likely to absorb city manners than the lower classes, not all city people shared an urban way of life.

Conversely, in many areas, villages or rural settlements differed less from towns or urban settlements than one might think. Places called villages by the geographers (presumably because of their relatively small size, agricultural orientation, and limited facilities) very often had pronounced urban features. In the frontier regions of

19 Yāqūt often reflects uncertainty as to whether places were properly quarters (*maḥalla*) or villages (*qarya*). See: Yāqūt, I, 135, 262; II, 112, 153, 237; III, 40, 391–92, 416; IV, 282, 300; VII, 164, 280. Villages were often found at the gates of cities, and were sometimes incorporated within the densely built-up area. See Yāqūt, I, 219; II, 164, 300; III, 43, 47–48, 117, 163, 167–69, 345; IV, 285; J. Sauvaget, "Décrets Mamelouks de Syrie," *Bulletin d'Études Orientales*, II (1932), p. 20. Conversely, villages at some distance might be regarded as quarters (Yāqūt, II, 164; V, 170).

Walled gardening suburbs are frequently mentioned (Barthold, pp. 104–105; Ibn Ḥawqal, II, 487–88, 509; al-Iṣṭakhrī, pp. 139–40, 185–86; al-*Idrīsī*, I, 497; II, 203, 205; Le Strange, *Eastern Caliphate*, pp. 274–75, 474; Nāṣir-i Khusrau, p. 136). In fact, so common must they have been, that Nāṣir-i Khusrau (p. 10) noted with surprise that the gardens of Kazvin were not walled. Gardens were also attached to palatial residences, or to houses generally, in not too densely built cities. See: Barthold, pp. 88, 171; C. Cahen, "La Djazira au milieu du treizième siècle d'après 'Izz-ad-Dīn Ibn Chaddad," *Revue des Études Islamiques*, VIII (1934), p. 113; Ibn Ḥawqal, I, 146, 221; II, 279, 493; al-Iṣṭakhrī, p. 117; al-*Idrīsī*, I, 464; Le Strange, *Eastern Caliphate*, pp. 95, 198, 303–05, 394, 396, 463–64; Le Strange, *Palestine*, pp. 375–76; al-Muqaddasī, pp. 320–21; Nāṣir-i Khusrau, pp. 28, 132–33; Schwarz, *Iran*, I, 52; Yāqūt, I, 63. In some cases, decayed old city centers were turned over to agriculture when commercial activity concentrated in former suburbs (Ibn Ḥawqal, I, 239; Le Strange, *Eastern Caliphate*, p. 62; Nāṣir-i Khusrau, p. 272). Even "city" mosques sometimes stood in the midst of gardens rather than in the actual city center (Le Strange, *Eastern Caliphate*, pp. 93, 111, 309; Nāṣir-i Khusrau, p. 25; al-Muqaddasī, p. 433; Yāqūt, VI, 332).

Tinnis and Damietta, which were entirely engaged in manufacturing and trade and not in agriculture, are noted as exceptional (Minorsky, Ḥudūd, pp. 151–52).

20 Baghdad: Ya'qūbī, p. 234. Fez: Le Tourneau, pp. 30, 33–34, 51. Cairo: S. Labīb, *Handelsgeschichte Ägyptens im Spatmittelalter [1171–1517]*, (Wiesbaden, 1965), p. 493. Damascus and Aleppo: Lapidus, pp. 84, 86–87, 90–91.

Khurasan where no substantial towns existed, villages were fortified to provide local security. More commonly, villages were sites of periodic markets and fairs. Others were caravan stations equipped with *khāns* or *ribāṭs* for travellers and sometimes permanent shops and bazaars. Villages were also centers of cloth manufacturing. Pre-modern industry, as evidenced by the "putting-out" system of European manufacturers, was not necessarily concentrated in towns, but was entrusted to villagers or peasants who earned their off-season income by spinning, weaving, and related activities. Neither were villages necessarily deprived of the spiritual facilities of towns. Some villages were sites of Sufi convents. Others had mosques, particularly Friday mosques, even though these are supposedly the distinguishing feature of town life. Many places called villages, which had mosques and markets, served the same functions as local marketing towns. In fact, though not in terminology, no distinction can be made between them. In Egypt, places called villages were sometimes the *chef-lieu*, residences of a governor or *qāḍī*, though elsewhere this seems to have been true only when towns decreased in size but retained their former jurisdictions. In short, these so-called villages had a full complement of urban facilities—baths, markets, and mosques.[21]

[21] Villages with fortifications: Bosworth, pp. 118, 159; Ibn Ḥawqal, II, 408–434; Ya'qūbī, p. 279. Fairs and markets: Barthold, p. 99; Ibn Ḥawqal, I, 217, 224; II, 351; al-Idrīsī, I, 316; Labīb, pp. 301–02; Le Strange, *Palestine*, p. 475; Schwarz, *Iran*, IV, 490; Yāqūt, IV, 103. Caravan halts: Ibn Ḥawqal, II, 349, 466; Ibn Rustah, p. 170; Le Strange, *Palestine*, pp. 467, 554; Yāqūt, II, 37, 357; III, 113; IV, 113; V, 322. Permanent bazaars: al-Idrīsī, II, 196; Yāqūt, II,262; III, 276, VII, 290.

Cloth manufacturing: Frye, *Narshakhī*, p. 16; al-Iṣṭakhrī, p. 93; al-*Idrīsī*, II, 148; Le Strange, *Eastern Caliphate*, p. 312; Le Strange, *Palestine*, p. 445; Minorsky, *Ḥudūd*, p. 110; al-Muqaddasī, pp. 202, 465; Schwarz, *Iran*, III, 238; Yāqūt, II, 40, 43–44; III, 299, 302–03; Ibn Serapion, p. 38.

Sufi convents: Yāqūt, I, 387–88; II, 440; IV, 110; V, 67; VII, 187. I did not think to accumulate references to shrines and graves of holy men, but they are also found in villages as well as towns.

Villages with a *jāmi'* or *minbar*: Barthold, p. 136; Frye, *Narshakhī*, p. 68; Ibn Ḥawqal, I, 132, 134, 142, 156, 172, 245; II, 372; Ibn Rustah, p. 179; al-Iṣṭakhrī, p. 60; al-*Idrīsī*, I, 350; Le Strange, *Palestine*, pp. 389, 436, 439, 471, 502, 509; al-Muqaddasī, pp. 77, 162, 176–77, 288, 317–18; Yāqūt, I, 214–15, 258; II, 128, 310, 312, 424; III, 12, 62, 105, 128, 454, 500; IV, 56, 60, 98, 225, 282; V, 185; VIII, 181. In fact, so much did the geographers take for granted that places called villages would have a *jāmi'* that they comment on its absence from many villages (Ibn Ḥawqal, I, 143; II, 257–58, 308, 313, 379, 498; al-Iṣṭakhrī, pp. 65, 99, 122, 180; al-*Idrīsī*, I, 387; II, 201; al-Muqaddasī, pp. 398, 409; Yāqūt, I, 200; Schwarz, *Iran*, IV, 356).

Villages with the facilities of marketing towns: Barthold, pp. 74, 98, 156, 158; Frye, *Narshakhī*, p. 12; Ibn Ḥawqal, I, 140–42; II, 363; al-*Idrīsī*, I, 477; Le Strange, *Palestine*, p. 537; Minorsky, *Ḥudūd*, pp. 136–37; Yāqūt, I, 318; II, 23, 324, 384, 403; III, 153; IV, 373.

Concomitantly, villages with varied activities had differentiated populations. Not only peasants, but landowners, *'ulamā'*, merchants, and artisans were also part of village populations. In Egypt, Khwarezm, the oasis of Damascus, and Bukhara, landowning families who were part of the cities' bourgeoisie resided in the villages. Villages were also the home of *qāḍīs*, preachers, *imāms*, scholars and holy men. They were resorts for townspeople, and in some special cases, as at Hormuz, merchants traded in the city but lived in the surburban villages.[22]

Such town-like villages were not exceptional. In regions with dense populations and highly developed agriculture—the Nile Valley and Delta, the Ghūṭa of Damascus, many parts of Palestine, the Jibāl, Khwarezm, and the important Eastern Iranian oases such as Bukhara, Nishapur, Merv, Rayy, Balkh, and Samarkand—the villages were *generally* as large in size and population, and as complex in facilities, as the places called *madīnas* or towns. These were "urbanized regions."[23]

Thus, many "cities" had rural components and villages had "urban" features or were in fact small towns in all but name. In some regions no hard and fast distinction between urban and rural habitats may be formulated. They form a continuum of geographical and ecological traits. To understand all the realities of geographical structure in the Muslim world, we should eschew the urban-rural dichotomy and avoid using "city" and "village" as absolute categories.

But if we abandon these familiar terms, how can we conceptualize the structure of settlement patterns? One possibility is to imagine settlements not in terms of towns and villages, but in terms of larger

---

Villages with political jurisdictions: Ibn Ḥawqal, I, 140, 142; al-Iṣṭakhrī, p. 135; Mustawfī, pp. 47–48, 68–69, 84; Yāqūt, III, 432.

Villages fully town-like: R. McC. Adams, *Land Behind Baghdad*, (Chicago, 1965), p. 94; Barthold, pp. 148–49; Ibn Ḥawqal, I, 138–42, 224; II, 362–63, 367; Mustawfī, p. 58. Some villages even had quarters and suburbs (Ibn Ḥawqal, I, 39; Yāqūt, I, 84; al-Iṣṭakhrī, p. 73).

22 Landowning families at Bukhara: Frye, *Narshakhī*, pp. 30–31, 53; Barthold, pp. 107–08. Damascus: Lapidus, p. 79. Egypt: Ibn Ḥawqal, I, 142.

Geographers of course give very little information on the lives of *'ulamā'* living in villages. A few examples are found: M. Barbier de Meynard, "Extraits de la chronique Persane d'Herat," *Journal Asiatique*, series 5, XVI (1860), 477–78; Frye, *Narshakhī*, p. 16, 139; al-*Idrīsī*, I, 332–33; Yāqūt, I, 214–15; II, 163; III, 106, 237, 405; IV, 59, 102; V, 185, 200–01; VIII, 434; Nāṣir-i Khusrau, p. 9. Biographical dictionaries should yield more information.

23 Ibn Ḥawqal, II, 379; al-Iṣṭakhrī, p. 123; al-Idrīsī, I, 350; Le Strange, *Eastern Caliphate*, pp. 216, 483; al-Muqaddasī, pp. 122, 155, 176; Yāqūt, I, 230; II, 55; IV, 356–57.

regional entities which encompassed both towns and villages. In many Iranian oases for example, whole regions may be imagined as composite "cities" in which the population was divided into non-contiguous, spatially isolated settlements. In the oases of Bukhara, Balkh, Bayhaq and Shash the population was settled in one dense city core, usually surrounded by suburbs, and in many small towns, villages, and hamlets scattered throughout the oasis. What made an entire oasis a single unit was the fact that not only the biggest settlements and their suburbs, but the entire region was surrounded by walls to protect against nomads. Because urban functions were not concentrated within the walls of the largest settlement, but were often distributed throughout the oasis, the outside walls, not the inside walls, may be conceived as the effective boundaries of the region. For example, a good deal of the manufacturing activity of Bukhara was not concentrated in the city core proper, but was dispersed in many small towns, villages, and estates. The Bukhara city proper was not always the residence of the governors and rulers, nor did the landowning elite, the *dihqāns*, live in the city proper before the tenth century; rather, they lived in villas scattered throughout the oasis. For good reasons, the name "Bukhara" applied both to the city proper and to the oasis region as a whole. Such oases may be regarded as extended boundary, multiple settlement composites.[24] Here the terms "urban-rural" or "city-country" tell us little about geographical realities.

Similarly, even though they were not bounded by walls, it would be realistic to consider many other oases or natural geographical districts composed of a number of larger and smaller closely connected settlements, and bounded by natural features, as regional composites. Also, the large cities surrounded by densely populated and closely interrelated suburbs and villages, can be considered in terms of encompassing regions rather than local divisions. To grasp all the realities of geographical settlement, we should think in terms of *pays*, districts, and regions, including both urban and rural units, as a natural form of settlement organization in the medieval Muslim world.

The notion of regional settlement composites denies neither the

[24] Balkh: Ya'qūbī, p. 288; Le Strange, *Eastern Caliphate*, pp. 420–22. Bayhaq: Bosworth, p. 260. Shash: Ibn Ḥawqal, II, 509; Le Strange, *Eastern Caliphate*, pp. 480–81. Samarkand: al-Iṣṭakhrī, p. 163. Bukhara: R. Frye, *Bukhara, The Medieval Achievement*, pp. 8, 23; Barthold, pp. 100, 102; Frye, *Narshakhī*, p. 34; Ibn Ḥawqal, II, 482; al-Iṣṭakhrī, p. 171; Le Strange, *Eastern Caliphate*, pp. 460–62; Minorsky, *Ḥudūd*, pp. 112–13.

distinctions between urban and rural habitats, nor the important differences between city and country people, nor the existence of a great variety of geographical and ecological relationships which prevailed in the Middle Eastern world. Rather it only attempts to formulate the meaning of hitherto neglected evidence bearing upon one important aspect of geographical organization, which is in turn related to corresponding forms of social organization. This view applies only to oases or other naturally bounded regions, and to highly urbanized areas with serious social, religious, economic, or political interactions among people living in various units. However, it does not apply to special cases in which cities were either isolated from their hinterlands or antagonistic to them, to areas where no substantial towns were found and in which small villages and hamlets predominated or to regions inhabited by nomads who did not share at all in settled life. However, despite many evident differences in the qualities of village and city life, there seem to be important resemblances between villages and towns in the class and occupational characteristics of populations, in the facilities available, and in the functions served. When such factors are taken into account, the contrast between town and village, or city and country, proves to be much less pronounced than we sometimes think. The distribution of population traits and functions does not necessarily support the feeling that cities *must* harbor a distinctive kind of society. Though many traits distinguish city people from country people, town populations as a whole have no discernible geographical or ecological quality which makes them single communities in contrast to rural dwellers.

THE MEANING OF *Madīna* AND *Jāmiʿ*

The conceptions of community organization and geographical form thus far developed pose an obvious difficulty. If communities were not defined by city spaces and often included both city and country dwelling populations, and if cities and villages were often similar in geographical and ecological qualities, how are we to understand the symbolic and religious significance which is apparently attached to cities in Muslim culture? If the Arabic word *madīna* means (as it is usually understood to mean) a city—a large settlement in which religious, political, economic and other functions are concentrated—how shall we reconcile our descriptions with the signification of the word? Furthermore, what is the meaning of the Friday mosque? If the *jāmiʿ* was both essential to the full realization of Muslim religious and social obligations, and was also a city institution, how can we reconcile the *jāmi'* with the concept of an Islamic community as jointly urban and rural?

Both problems may be solved, I think, by a close examination of the meaning of the terms. First, the word *madīna*, as used by the Muslim geographers, had no reference to the size or population of a settlement, nor to the complexity of its social or economic functions, nor to its differentiation from "rural" environments. *Madīna* was a word applied to places which varied in geographical scale and functional complexity from village-like to metropolis-like settlements, but not to every settlement in this range. Many villages were larger than places called *madīnas*. They were towns then, in the physical sense, but did not bear the name *madīna* because they lacked the essential attributes of a *madīna*, which were not geographical but jurisdictional. A *madīna*, in principle, was an administrative capital, often the capital of a *nāḥiyya* or *rustāq* district. Thus, for example, the geographers point out that Qinnisrīn was the *madīna* of its province even though Aleppo was the most important economic, cultural, and even governmental center. *Madīna* was a term which defined the position of a settlement in an administrative hierarchy.[25]

Furthermore, it seems that a *madīna* was always a religious capital as well. *Jāmi's* or *minbars* were always found in *madīnas*, but the converse was not necessarily true. If *madīnas* always had *jāmi's*, not all places with *jāmi's* were necessarily *madīnas*. *Jāmi's* were found not only in innumerable villages, but also in fortresses (*ḥuṣn*), in caravan way-stations (*manzal*), and sometimes in provincial districts which were not, so far as we can tell, *madīnas*, but which did have *minbars*. Religious centers could thus exist apart from political capitals. *Minbars* or *jāmi's* were found in places which were neither cities in the physical sense nor capitals in the political sense. The Friday mosque was a feature of settlements of all sorts, "urban" and "rural" from the geographical point of view, *madīna* and non-*madīna* from the political point of view.[26]

Thus, in fact, the *jāmi'* was neither a strictly urban phenomenon, nor, from the point of view of Muslim legal theorists, was it supposed to be located in *madīnas*. In the *al-Aḥkām al-Sulṭāniyya* by al-Māwardī, it appears that the Muslim law schools were divided as to whether or not a *jāmi'* required an urban location. Al-Māwardī

[25] For villages with attributes of towns, see above. Qinnisrīn: al-Iṣṭakhrī, p. 49.

[26] *Minbars* in villages: see above. Fortresses: al-Iṣṭakhrī, p. 47. Way stations: al-Iṣṭakhrī, p. 159. *Rustāq*: al-Iṣṭakhrī, pp. 69–71; Ibn Ḥawqal, II, 266.

points out that according to the Shāfi'ī school, the *khuṭba* may be said in any permanently inhabited place, whether it be a village or a city, while according to the Ḥanafī school, it must be restricted to cities. Al-Māwardī was himself a Shāfi'ī, and seems to state the former view as his own. Judging from the geographers, the Ḥanafī restrictions were not in force from the tenth century on, and the Shāfi'ī view represented and validated the actual situation. In fact, and in the theory of at least one very important school, the *jāmi'* and the pronouncement of the *khuṭba* had no intrinsic relation to town settlements or *madīnas*.[27]

To identify *madīna, jāmi'* and city (in the physical sense) may be an adequate approximation because the three aspects so often went together, but it is an approximation which obscures the actual meaning of the concepts and cannot serve to account for anomalies. The *madīna* was not a "city" but any place with political (and usually religious) jurisdictional supremacy. Nor was a place with a *jāmi'* necessarily either a city or a *madīna*. Correspondingly, places with the geographical and ecological attributes of small towns did not necessarily have a Friday mosque or the status of *madīna*. In Muslim usage, geographical city, political capital, and religious center must be analytically distinguished.

If *madīna* and *jāmi'* may be separated, what rules governed the placing of Friday mosques? To this question only a very incomplete and tentative approach may be suggested.

First, as Professors Lassner and Grabar have pointed out, the early centuries of Islam must be distinguished from the later period. Until the middle of the tenth century, Islam was a political-*cum*-religious community governed by its caliphs. They were responsible for communal worship, and communal worship was not fully valid without the presence of the caliph or his representative. Mosque and *madīna* seemed indissolubly attached. Moreover, until the tenth century, permission of the caliph was required for the founding of a mosque and such permission was only rarely accorded. The mosque was not intended to serve everyday religious needs, but to symbolize the religious solidarity of the community as a whole as formed by the organizing authority of the caliphs, successors to the Prophet and administrators of the divine law.

Only late in the tenth century did mosques begin to multiply within Muslim settlements. By then the largest cities—Baghdad,

27 al-Māwardī, *al-Aḥkām al-Sulṭāniyya*, tr. E. Fagnan, (Paris, 1915), pp. 209 ff.

Basra, Nishapur, Merv, Cairo, Alexandria, and others—had two mosques, possibly to provide facilities in proportion to population, but more probably reflecting complex considerations in the formation of Muslim religious life.[28] Already by the end of the tenth century the Ḥanafī *'ulamā'* of Transoxania were resisting the proliferation of new *jāmi's*. The *'ulamā'* of Fārāb objected to the founding of a new Friday mosque in the suburb of Kadar, and the *imāms* of Bukhara closed a Friday mosque in Shargh. The geographer al-Muqaddasī remarks on how difficult it was to establish new *jāmi's* in villages even though they already had all of the attributes of towns.[29] Evidently, important religious principles and jurisdictional considerations were involved. We may speculate that such disputes were simply quarrels over the distribution of authority, income, and other perquisites, but unfortunately we know too little about the early schools of law to specify precisely what was at issue. For the present, we may suppose that resistance to new *jāmi's* and more importantly, their eventual multiplication on a massive scale in the eleventh and twelfth centuries, reflects the evolution of the sort of communal organization I have tried to describe. The *jāmi'* was less and less directly attached to the state and more and more attached to local religious communities. For example, the *'ulamā'* of Nishapur and Merv were expressly affiliated with particular *jāmi's*, and in cities with several schools of law or several sects, *jāmi's* were founded for each main grouping. Bamm in Kerman, in the tenth century, had a Friday mosque for the Khārijīs and another for the Sunnis. From the twelfth century on, Mesopotamian, Syrian and Egyptian cities all had religious facilities for each of the schools of law. In the course of time, the *jāmi'* came to be more expressly identified with religious communities and less identified with political hierarchy. *Madīna* and *jāmi'* were no longer equated. Mosques, though essential to Islam, did not define a city community as opposed to other communities.

Thus, in principle the community served by the *jami'* was not the "city" but a specific religious grouping formed without respect to the precise space in which the communicants lived. In the composite settlements we have already described, on both a geographical and social level, the *jāmi'* served as a community center for people

[28] Ibn Ḥawqal I, 144, 147, 216; al-Iṣṭakhrī, pp. 118, 147; al-Muqaddasī, pp. 117, 197, 310–12.

[29] al-Muqaddasī, p. 282; J. Lassner, *The Topography of Baghdad*, unpublished dissertation; Frye, *Narshakhī*, pp. 14–15; Barthold, pp. 99, 120, 176.

living within the towns or villages where it was located and for people dispersed in the surrounding region. Thus, for example, the villagers around Ludd in Palestine came there for the *khuṭba*. In the Oxus region, the people of Akhsīsa went to Zamm.[30] The *jāmi'* signified not the existence of a *madīna* or city (in principle, though in many cases they overlapped), but the organization of a larger religious community transcending the confines of particular settlements. The mosque did not divide city and country, but united different settlements into larger religious communities. Townsmen might have been more pious or more learned, but in principle Islam made no distinction between the religious dignity of city and country people any more than it made caste, sacerdotal, political or other differentiations of religious worth.

CONCLUSION: "CITIES" AND ISLAMIC SOCIETIES

Despite the temptations of "common sense" and the pressure of historical and cultural ideologies, we can no longer think of Muslim cities as unique, bounded, or self-contained entities. From the study of social organization we find that none of the characteristic social bodies of Muslim society—the quarter, the fraternity, the religious community, and the state—were specifically urban forms of organization. Nor were any of the groups identified with cities *qua* physical settlements. Muslim populations were organized into groups which formed subcommunities within city spaces and super-communities of religion or state which extended beyond any single city space. Cities, in this view, were simply the geographical locus of groups whose membership and activities were either smaller than or larger than themselves. Cities were nodules of population woven into the fabric of a larger society—places which concentrated persons and activities, facilitated the organization of populations which cut across their own space, and helped them to resolve their relations to one another. Cities were physical entities but not unified social bodies defined by characteristically Muslim qualities.

Moreover, Muslim cities were no more isolated as physical bodies than as social bodies from the larger world in which they were embedded. Muslim settlements formed geographical and ecological, as well as social composites including territories and populations which were neither exclusively urban nor exclusively rural, but a

[30] Bamm: Ibn Ḥawqal II, 312; al-Iṣṭakhrī, p. 99; Minorsky, *Ḥudūd*, p. 125; al-Muqaddasī, p. 469. Nishapur: Bulliet, *Nishapur*. Merv: al-Muqaddasī, pp. 310–12. Ludd: al-Muqaddasī, p. 176. Akhsīsa: Le Strange, *Eastern Caliphate*, p. 443.

combination of the two. The integration of villages, quarters, suburbs and towns into larger geographical units provided a physical context for the kind of social pattern we have described. Furthermore, the role of the *jāmiʿ* in Muslim religious life seems entirely consistent with the actualities of social and geographical organization. From this point of view we should not speak of "Muslim cities," but of settlements in the Middle East harboring a certain kind of Muslim society. If these hypotheses are well founded, we may begin to supersede our interest in cities as a crucial topic of historical, sociological and cultural investigations by a more differentiated conception of the relationships between social structures, geographical forms, and religious meanings.

DISCUSSION

Professor Goitein broached the question of the relationships between cities and villages: "What after all is a city? *Madīna* is an Aramaic word and *dīn* means justice. *Madīna* means first of all a place where justice is administered, where the government or its important representatives have their seats, and where there is security—a larger amount of security than elsewhere. Villagers came on Friday to the city in order to attend the market and to see the government." Professor Goitein stressed the "eternal contrast between the town dweller and the *fallāḥ*—the man who gets up at sunrise and tills the soil until the sun goes down. This is a very different type of man from the man who lives in the city. The discrepancies between city and country are great. For example, an hour out of Jerusalem the people speak an entirely different Arabic from the Arabic which is spoken in the city. Take taxation which is so important. Who was taxed? Who had to bear the expenses for the state? The *fallāḥ*, the farmer. Commerce was taxed, but to a far smaller extent.

"It is true that the aristocracy was landed; everyone had land in addition to his place in the city. But in the Geniza two types of farmers are indicated. One is the land owner who grows crops which are sold for money, such as flax and indigo. He was a respectable man and could be related to the families in the town. But the *fallāḥ*, the peasant, was quite a different type of person.

"My own research in earlier years has convinced me of this difference. In 1949 I went to Yemen and I made a catalogue of 1,050 little village communities. There were many villages in which highly cultured persons lived. But other people had lower standards. These were simple folk, according to our concept, who were not articulate. So I think there are very deep and basic differences."

In another vein, Professor Grabar highly valued the emotional or belief component in defining cities: "It is essential to consider the degree to which there was an emotional attachment to cities, for valid or invalid reasons, and to explore the nature of this attachment, and what unique qualities of living in a city as opposed to a village, or in one city as opposed to another, this attachment implies." Professor Goitein added that "this kind of attachment was felt by Muslims who wrote histories of the scholars who lived in or visited particular cities. People were proud of their cities."

Professor Adams inquired further about the significance and the appropriate interpretation of the word *qarya*, the term used by the Arabic geographers for village. Some of his personal experience suggests that it is a relativistic concept. "In surveys of pre-Islamic settlements east of Baghdad, I have been struck with the fact that the importance of towns, as can be inferred from their size in given periods and from their central position in relation to major irrigation works, has no correspondence to their appearance in the works of the Islamic geographers of the tenth century. This suggests to me that what we really find in the texts are the communities, whatever size they may be, that happen to be important in relation to the city because they lie on routes or are important toll stations on the main waterways. Therefore, as seen from Baghdad in this case, they are important enough to rate mention as *madīnas*, whereas much larger and functionally more important settlements are simply neglected and quite possibly spoken of as *qarya*."

Professor Abu-Lughod noted, referring to the contemporary situation that the statistical distinction in Islamic countries between urban and rural often has little to do with the demographer's distinction: "An urban community may be small, may be rural, and the population may be engaged in agriculture. The distinction is an administrative one. Thus, when one examines census data for an urban-rural breakdown, one may find communities classified as urban solely because of their administrative roles, while conversely, some large cities do not play an administrative role and therefore are not classified as urban. Conversely, there are some very small agricultural communities with administrative roles which are classified as urban."

Professor Fernea raised an important question as to "the sense in which a religious community may be said to link rural areas and urban areas. Professor Lewis has observed that in the medieval period, the commitment to orthodox Sunnism was very much limited to an urban elite and heterodoxy of various kinds prevailed even within the cities. Although people might have been nominally Muslims of Sunni persuasion and may have come into the community for juridical purposes, how deep do such religious commitments actually go?"

Professor von Grunebaum commented on the complexities of the comparative problem: "The Greek experience gives us the model of the *polis*, but the Greek model city tends to blur the distinction between the city—the walled *polis*—and the region around it. In the same way in which Dr. Lapidus describes Hamadan, Athens had reserved for itself full jurisdiction. What voting the villagers did, they did in the town. The movements between the village and the town were constant, and the town functioned as a protecting agency, as a centralized administrative structure. The villages were in no sense distinct as political jurisdictions or economic entities from the town itself. Still, some differentiation of dialects could be found. People who came from the villages were ridiculed for their un-Attic dialect. What establishes the main similarities, from the legal point of view, is that the rural land owners, who were all Athenian citizens, did not have the right of independent jurisdiction over their villagers or the places where they had holdings. Although this is exactly the same in the Muslim case, it is in contradistinction to what happened in the Byzantine and European societies."

Professor Safran took up the comparative problem in its political dimension and asked, "Why is it that the Muslim city did not develop the corporate political entity which played such a central role in the political development of Western Europe? Dr. Lapidus suggested that the Muslim city is too diffused to have any dominant function or dominant interest, while Mr. Goitein suggested that there is, as the name *madīna* suggested, a dominant function: this is the place where the government is located. I would suggest that these two ideas are related. There were no strong interest groups in the Muslim cities as there were in the medieval Western cities where the burgher groups, for example, played such an important role. There were no such interest groups partly because the *madīna* was the site of the government and was dominated by the government. If we look at the development pattern of medieval Western cities we can see how this circular thing fits. Western cities developed in a feudal society in which the main center of power was based on the land—on the estate. They developed in a completely agricultural society with a political structure that was fitted to it. Cities developed on the fringes of the main centers of power in that society, and were able to grow to a point which enabled them to challenge the centers of power. They had to develop their own laws and regulations. However, in Islam, where cities had developed continuously since antiquity without the break that occurred in Western Europe, where cities had always been seats of government, and where the government as the dominant urban interest did not allow other interests to develop, no independent society emerged. There were no autonomous laws and regulations in Muslim cities, and hence there were no corporate interest groups challenging the caliphate and forcing the development of constitutional law, as took place in the West."

Professor Grabar added that there were attempts made to develop strong local government and constitutions but they failed.

Professor Moore saw the situation of Fez in a similar light: "Fez never had a municipal government; there was not even a municipal budget, for the management of various public services were performed by a variety of groups which were not even distinctively urban. Apparently even the eighteen quarters into which Fez was divided performed few governmental functions. The *muqaddam*, or official appointed by the city's governor after nomination by the notables of his quarter, was not himself a notable—one index of the slight importance leading urban families attached to this office. There was enough communication between the quarters (indeed, the real unit of local solidarity was not the heterogeneous quarter but rather the neighborhood defined by alley ways and impasses), so that notables of the city could identify one another; they could more or less agree on a complex set of criteria for defining the true "Fassi." In a sense the Fassi notables formed a vaguely defined community, but the sense of social solidarity acquired political significance only in times of crisis, as during succession crises or periods of tribal revolt, when the urban militia would defend the city against rampaging tribesmen. Like other potential communities in Moroccan political life, the city was an intermittent political system; its sense of community, awakened only at times by external threat, was never sustained by a political organization through which autonomous urban interests could be articulated and public discourse fostered about common concerns.

"Furthermore, I think that the absence of urban political community accounts in part for the relative ineffectiveness of traditional urban elites in twentieth century North Africa and possibly elsewhere. Muslim urban notables did not form a status group distinct from the various ascriptive groups to which they belonged; they could not serve as the core for articulating a public interest, or for mobilizing new groups into a cohesive nationalist force. Having virtually no concrete political experience, the urban notables substituted rhetoric for sustained organizational efforts. From this stems, in part at least, the political fragmentation of contemporary Morocco. In Tunisia on the other hand, political cohesion has been possible because nationalism triumphed only after a generation of French-schooled sons of Sāḥil peasants displaced the old Tunis elite in the 1930's as the core group. Nationalism became an effective force when whole regions (perhaps akin to the Iranian oases mentioned by Professor Lapidus) were transformed into a political community through the agency of party organization."

Professor Adams remarked: "I'd like to pick up the very illuminating contrast that Professor Safran was developing and suggest a slight modification. Rather than seeing city-states as developing purely on the peripheries and rather than seeing continuity as being the dominant characteristic in the Middle East, I think it is perhaps worth considering the case of Iran and the very close connection between strong extended imperial domains and the founding of cities. This is particularly true just before Islamic times, when the Sassanian conquerors founded new major cities over wide areas of their realms. I think this has to be understood not simply as a royal whim, but as an attempt to fundamentally alter, in the interest of the central power, the relationship of the ruler to the nobility. By bringing into being new systems of irrigation, founding new cities, forcing shifts in population into areas which were firmly under royal command, and giving out land to royal favorites, the central government strengthened itself. Here cities were involved within the polity in a very different way."

Turning to other matters, Professor Grabar was not certain that it is proper to call a village mosque a *jāmi'*, except "colloquially." He observed that if the development suggested in his paper beginning about 1100 A.D. is correct, "there would have been a multiplication of religious buildings. By the twelfth century a place that merely had a mosque, even with a *minbar*, did not necessarily qualify as a city. What qualified as a city was a place that had the whole range of different kinds of religious buildings—*madrasas, ribāṭs*, etc. In this situation, a place that only had a *minbar* or *masjid* could still be a village, for the mosque or *minbar* had lost its significance as a defining characteristic of cities.

"There is a series of curious examples in the twelfth and thirteenth century towns in the *Jazīra*, or what is now primarily south-eastern Turkey. Some of these towns were created suddenly because of the need to fight against the crusaders, or because of the so-called Muslim feudal system. When a prince took over a conveniently located place, he developed a town there. Often the town did not exist longer than this prince or his immediate successors, and again became a small village. In many of these cases a prince built a citadel and then a mosque, a large mosque which remained as the village mosque ever after. Although there may be monumental buildings in secondary cities and secondary villages, they were created under the assumption that the city would continue to have a greater importance than it ever actually maintained."

Professor Lapidus added that "*jāmi'*s were widely diffused in villages by this period, and that the introduction of *jāmi'*s in villages goes back at least as far as the tenth century according to the geographers."

S. D. GOITEIN

Cairo: An Islamic City in the Light of the Geniza Documents

The Cairo Geniza throws new light on Fusṭāṭ, the ancient capital of Islamic Egypt.[1] Geniza (pronounced gheneeza) is a lumber room in which discarded Hebrew writings used to be put away. The term "Cairo Geniza documents" refers to material dating mostly from the tenth through the thirteenth centuries, written mostly in Hebrew characters but in Arabic language, and originally preserved in a synagogue, and partly also in a cemetery of Fusṭāṭ. The material comprises every conceivable type of writing, such as official, business, and private correspondence, detailed court records and other legal documents, contracts, accounts, checks, receipts and inventories, writs of marriage, divorce and manumission, prescriptions, charms, and the like. These writings originated mostly in the middle and lower classes and are therefore invaluable for the knowledge of social groups to which the historian until now, has had little or no access.

The Geniza contains several hundreds of papers in Arabic script, partly emanating from Muslims or Christians. But the bulk of the material is of Jewish origin, naturally, and there arises the question of how far it may be used for a description of the Islamic city in general.

To answer this question, we have to bear in mind that no ghetto existed in Fusṭāṭ. In most of the contracts related to houses we find

[1] This study is almost entirely based on unedited manuscripts from the Cairo Geniza, preserved at the University Library, Cambridge, England, and many other libraries. About these collections and the Cairo Geniza in general, see S. D. Goitein, *A Mediterranean Society: the Jewish Communities of the Arab World as Portrayed in the Documents of the Cairo Geniza,* vol. I: *Economic Foundations,* (University of California Press, Berkeley and Los Angeles, 1967), Introduction, pp. 1–28.

For published Geniza documents see S. Shaked, *A Tentative Bibliography of Geniza Documents,* (Paris and the Hague, 1964).

For the topography of Fusṭāṭ see P. Casanova, "Essai de reconstruction de la ville d'al Fousṭāṭ ou Miṣr," *Mémoirs de l'Institut Français d'Archéologie Orientale du Caire,* XXXV (Cairo, 1913–1919).

that houses belonging to Jews bordered on Muslim and/or Christian properties. Muslims lived in houses belonging to Jews and vice versa. Nor was there an occupational ghetto. Jews were prominent in certain industries, such as gold- and silversmithing, textiles (especially silk, and dyeing) as well as glassblowing. In business they were engaged in the corresponding trades and, in particular, also in pharmaceuticals; but here, too, there were no watertight compartments. We find partnerships between Muslims and Jews in both industry and commerce, and there were many other ways of cooperation as well. The same urban taxes were paid by Muslims and non-Muslims. Still, the specific character of the environment in which the Geniza documents originated has to be taken into consideration when they are used for historical research.

As far as the topic of this conference is concerned, we are in the happy position of being able to countercheck the Geniza material from many different quarters, for our information about the Islamic city during the High Middle Ages is rich and variegated. The literary sources are abundant. Some buildings erected at that time still stand; many other artifacts such as inscriptions, coins, utensils, ornaments, and textiles also exist today. In dealing with the Geniza I shall confine myself to describing the city as a place of human habitation. Moreover, I shall not make an effort to arrive at generalizations and comparative conclusions, but shall present the information gathered directly from the documents. I shall act not as a *sociologist* but as a *sociographer*.

The first Arab Islamic cities were founded and filled with populace almost immediately after, or even in the course of, the conquest and formed the nucleus of Arabization. Though Arabic and Islam were politically predominant, it took hundreds of years until the villages assimilated the religion, the language and the social traditions of the conquerors. A country used to be identified by the Arabs with its capital city, which at the beginning was the only place where Arabs lived in great numbers, with the result that the capital used to be called by the name of the country. Thus the official name of Fusṭāṭ, the capital of Egypt, was Fusṭāṭ Miṣr, "Fusṭāṭ of Egypt," but throughout the Geniza papers, except in legal documents, it is spoken of as Miṣr, "Egypt." Damascus is referred to as Shām, "Syria." Palermo, the capital of Sicily, is usually called Siqilliyya in the Geniza papers, sometimes also Madīnat Siqilliyya, "The City of Sicily," abbreviated al-Madīna, "The City," while Balarm, the Arabic equivalent of Palermo, is virtually absent.

In Roman times, the provincial capitals, such as Caesarea of Palestine, Alexandria of Egypt and Carthage of the province of Africa, bordered the sea where they could be reached directly from Rome, the center of the world. Under Islam, the cities chosen or founded as capitals—Fusṭāṭ, Damascus, Kairouan—were situated inland, where they were safe from surprise attacks by the Byzantine and other Christian navies. The great ports, such as Alexandria, Tyre on the Lebanese coast, or al-Mahdiyya were regarded as frontier fortresses. Alexandria is often referred to in the Geniza letters not by its name but as *al-thaghr*, "The Frontier Fortress."

It would be wrong to assume that Alexandria, the Mediterranean port, was the pivot of commerce, while Fusṭāṭ, originally the capital of the country, derived its riches from its function as seat of the administration. The unmistakable testimony of hundreds of Geniza letters proves that Fusṭāṭ, although inland, was also the commercial and financial center on which Alexandria, the originally Greek maritime town, was economically dependent in every respect. Fusṭāṭ was the emporium of the region, where all the goods were stored. Even commodities from Mediterranean countries, which were imported by way of Alexandria, had to be obtained from Fusṭāṭ when they became scarce in the former city. Foreign currency from places with which Alexandria entertained a lively commerce was regularly acquired by Alexandrians in Fusṭāṭ. Humdrum commodities, such as shoes and clothing, implements for silversmithing, or parchment were ordered from Alexandria in the capital, "for here, [namely in Alexandria, as several letters emphasize] nothing is to be had." To be sure, Alexandria, like other places, had its local industries, such as its world-renowned mats, which were popular everywhere, its *maqṭaʿ* textiles, exported as far as India, and its pens, cut from the reeds of Lake Maryut. It seems also that it served as the international entrepot and bourse of the silk trade.

Besides the inland capitals and their port cities, there were a number of other urban centers of considerable size whose names constantly recur in the Geniza papers. Two queries submitted to Maimonides take it for granted that the minimum requirements for an urban place were a synagogue and a bathhouse. This echoes, of course, the Muslim definition of a city as a place with a Friday mosque, a market and a bathhouse. Quite a few localities in Egypt were of this type. However, for the inveterate city dweller everything outside the capital and Alexandria was *Rīf*, a term insufficiently rendered as "countryside" since it included towns of no small extent.

Many marriage contracts stipulated that the husband was not permitted to exchange the city for another place except with the consent of his wife. In one letter we read that a woman could not bear "life in the *Rīf*" although she was married in the provincial capital Mahalla, an important industrial town then as now. She escaped to Cairo, but her husband declared that he could manage at most to gain a livelihood in Damietta, where he offered to move in order to please his recalcitrant wife. A retarding debtor would abscond from the city and try to hide in one of the provincial communities, wherefrom we find a circular from a *nagid* addressed to "all the judges and *muqaddams* in the *Rīf*" to get hold of him. Exigencies of business or office forced many a Cairene or Alexandrian to take up his domicile temporarily in a smaller town, and many a bitter complaint about the "wrongs" suffered there, about the "Sodom and Ghomorrha" endured from its uncouth people, may be read in the Geniza letters. That a member of the urban *jeunesse dorée* appointed to posts in the *Rīf* again and again forsook them for the city is, of course, not surprising. Community officials implored their superiors to transfer them to the city (usually with the rationalization that they might continue their studies), and businessmen would come to the conclusion that after all it was wiser to stay home in the city than to seek profit in the *Rīf*. The idea of being forced to pass a holiday "outside," meaning outside the capital or Alexandria, was particularly abhorred. The incessant exodus from the *Rīf* to the cities is illustrated by the great number of persons bearing family names derived from small and even obscure places in the Egyptian countryside.

Despite this contrast, or even tension, between city and *Rīf*, the Geniza papers reveal the importance of the smaller towns. There was not only a tendency to move from the *Rīf* to the cities, but also one in the opposite direction. Persons bearing the family names of Miṣrī, Iskandarī, 'Asqalānī, Qudsī or Dimashqī, i.e., originating from Fusṭāṭ, Alexandria, Ascalon, Jerusalem or Damascus, respectively, are met with in the small towns of Egypt. In fact, we find there people from all over the Mediterranean, including Byzantium and France.

In more than one respect the small towns were of vital economic importance for the people who left us their records in the Geniza. The staple exports of Egypt, flax and indigo, needed supervision at the source and were also partly processed where they were grown. The manufacture of cheese, the main foodstuff after bread, was preferably done where the best sheep were bred and had to be supervised since it was subject to ritual taboos. Jews, although to a lesser degree than Christians, were widely active as tax farmers and as such had to live in small towns or even villages in order to collect the taxes in person. They were also employed by the government in the *Rīf* in other capacities. Jewish physicians and apothecaries seem to have been ubiquitous in the *Rīf*. The same was perhaps true for the Jewish silversmiths, dyers, and workers in silk. The economic strength of the *Rīf* is demonstrated also by the fact that it was regularly visited by traveling merchants and agents selling their silk and other textiles, as well as by people soliciting funds for works of public or private philanthropy.

Still, most of the data on urban life refer to the capital, and in the following, wherever no remark to the contrary is made, Fusṭāṭ is meant. Fusṭāṭ does not mean Cairo, for, although the twin cities were separated from each other by only about two miles, they largely lived distinct lives. The marriage contract of a wealthy bride from the year 1156 stipulates: "She will live in Fusṭāṭ [called, of course, Miṣr in the document] to the exclusion of Cairo." No reason is given for this condition, but it demonstrates how strongly the contrast was felt almost two hundred years after the foundation of Cairo. The Maghrabī traveler Ibn Sa'īd, who visited Egypt in the 1240s, describes the people of Fusṭāṭ as soft-spoken and far friendlier than those of Cairo, and expresses his astonishment that there should be such a difference between two adjacent cities. The reason for the contrast probably was the fact that Fusṭāṭ was dominated by an easygoing middle class, while Cairo lived under the shadow of a stiff court. Originally, Cairo "was the seat of the caliphate and no one lived there except the caliph, his troops, his entourage, and those whom he honored to be near to him." There is indeed documentary evidence that the Fāṭimid court physicians, higher government officials and other persons connected with the court lived there. When a Jewish government official was fired from his post, he was ordered to leave Cairo and to take up residence in Fusṭāṭ. The Gaon, *Maslīaḥ*, or head of the Palestinean academy, who moved to Egypt in 1127, bore the titles *Jalāl al-Mulk* and *Tāj al-Riyāsa* ("The Luster of the Empire" and "The Crown of the Leadership"), although he was neither a physician nor an official. This, together with the robes of honor bestowed on him according to an Arabic Geniza letter, indicates that he belonged to the caliph's retinue. We find indeed that he and some of the scholars attached to him lived in Cairo. Others had houses both in Cairo and in Fusṭāṭ.

In the Geniza, most paradoxically, Cairo—especially in Fāṭimid times—is represented not by the upper crust, for which it was renowned, but by a social layer which was both materially and spiritually poorer than many a community in a provincial town. The houses and dowries mentioned in the documents are of low value; the documents themselves are often of poor quality and signed by persons whose handwriting betrays them as having had little practice in writing. This section of the population provided the menials, such as water carriers, or consisted of destitute persons who had come to the city seeking help. A young Karaite, who implored his fugitive wife in the most humble terms to return to him since it was impossible for a chaste man to live a bachelor's life in a city like Cairo, belonged to the same stratum of the society. This anomaly is probably due to the fact that persons connected with the court presumably corresponded in Arabic characters, wherefore they had no reason to discard their letters in the Geniza chamber; or they discarded their Hebrew writings in their own synagogue which has long since disappeared.

Trying now to describe the main features of a city of the type of Fusṭāṭ as they appear in the Geniza documents, we have first to recall a few facts about the history of that place. The nucleus of the city was formed by a Byzantine fortress, situated on the eastern bank of the Nile and called The Fortress of the Greeks or The Fortress of the Candles. The Christian and Jewish elements remained preponderant in this pre-Islamic section of the city, and time-honored churches, some of which are still extant, as well as the two main synagogues were found there. Some of the towers of the fortress were used as houses and workshops and partly were owned by the synagogues as pious foundations. The Islamic city, founded at the time of the conquest (641 A.D.), surrounded the Fortress of the Greeks from the north, east and south, and later, when the Nile silted up, also from the west. It was laid out according to the tribes, clans and groups which composed the conquering army. The area reserved for the High Command and its guards, extending from the Fortress of the Greeks north-eastwards to the newly founded mosque of 'Amr, the conqueror, was called *Ahl al-Rāya*, "Those around the standard" (the flag), or shortly *al-Rāya*. Our documents reveal the interesting facts that six hundred years after the Muslim conquest the main quarters were still being called by the names of those ancient Arab groups, such as Banū Wā'il, Wa'lān, Khawlān, Tujīb, Mahra, and Banāna, and that al-Rāya, the original center of the Muslim city, was, in Fāṭimid times, largely inhabited by Christians who were in the process of being bought out by Jews.

A great many localities in Fusṭāṭ are referred to in the Geniza documents; but, as far as we are able to discern, with a few exceptions all were situated within a perimeter at most one mile long (northeast-southwest) and 700 yards broad. This area must have been closely built up. At least half of the streets, lanes and alleys referred to are described as having no outlet, and the rich topographic vocabulary of the documents does not contain a word for public square, which can only mean that there was none. But open spaces were not entirely absent from the city. There were cattle pens, called *zarība* (cf. English *zareba*), where the cows and sheep were kept which provided the daily milk for the population. Repeatedly we read about a *misṭāḥ*, or place for spreading things, presumably dyed cloth. Promenades and parks were not lacking in the medieval capital of Egypt. The vizier al-Ma'mūn (1121–1125) was credited with having created or re-established not less than five of them. But all of them lay outside the perimeter alluded to above. When the *nagid* Abraham Maimonides asked a man of his entourage to slaughter for him a sheep "at the little garden," *junayna* (in case meat needed at a reception could not be had on the market), he meant perhaps the "orchard" of the Synagogue of the Palestinians, although this place, like the gardens in private houses, was usually referred to as *bustān*. That a rose garden should be found in the midst of the city is not as strange as it might appear in view of its otherwise so cramped quarters. Rose-water was one of the most popular medical potions, wherefore the growing of roses on the expensive ground of the city was economically justified.

Was there any zoning, i.e., compulsory division of the city into residential, commercial, and industrial quarters? That some such division existed is evident from the very names of many bazaars, markets, squares, streets and other localities—names which indicate a specialization in a certain trade or industry. On the other hand, the Geniza proves that this specialization was not too strict; for example, a street of cobblers could harbor the shop of a perfumer. In many cases, houses forming the object of deeds of transfer must have been situated in quarters exclusively or predominantly residential since the properties forming their boundaries are described as the homes of private persons. At the same time, we find high-priced houses, obviously serving as homes, bordering on busy bazaars, and we find stores topped by an *'alw*, i.e. one or several upper stories. Stores are sometimes referred to as living quarters. We find a physician living on a street of waxmakers and a judge in a bazaar of druggists. A sugar factory belonging to a physician is expressly described as also being his domicile. Conversely, a petition with regard to the estate of an industrialist clearly discerns between the family's *maṭbakh*, or factory, and *sukn*, or living quarters. An interesting, but very much mutilated, letter says: "People who had been living on their properties gave them up. You will sell the house and they will convert it into a workshop, *ma'mal*." The tenor of the letter seems to indicate that such changes were regarded as unusual. All in all it appears that some division into essentially residential and preponderantly commercial and industrial zones existed, but was not strictly enforced. Its enforcement was left to private initiative, both Islamic and Jewish law protecting a proprietor against any changes effected by neighbors which might be recognized as being harmful to his own property.

Steven Runciman tells us that in Byzantine Constantinople the slums of the poor jostled against the palaces of the rich. Our analysis of the prices of properties in Fusṭāṭ seems to show rather that, like our own cities, the Egyptian capital was divided into neighborhoods of homes of higher and lower values. In one respect, however, Fusṭāṭ and other Islamic cities must have represented a sight very much different from what we are accustomed to: everywhere homes, even newly erected ones, were interspersed with ruins. The abundance of references to ruins in the writings of the Muslim antiquarians of the fifteenth century could be explained as evidence for the general trend of decay typical for the late Middle Ages. However, ruins are as frequent in the Geniza documents of the twelfth century as in the pages of Ibn Duqmāq, and are as typical of affluent neighborhoods as they are of poor environments. Ruins are frequently mentioned as being adjacent to the homes of well-to-do people, either given together with them in a will, or contrarily, expressly excluded from a transaction. Very often they appear as a boundary, and once as a place where the proprietor was entitled to dispose of his rubbish. In view of this situation it is natural to find some persons rebuilding a ruin adjoining their house, others buying ruins and giving them as a gift or rebuilding a ruin purchased, and others again taking illegal possession of a ruin or parts of it. Finally, we read about people living in ruins and paying rent for them. The very usage of describing a house in a deed of sale as "built up" points to the fact that houses often were partly or totally in ruins. We shall presently discuss the economic aspects of this phenomenon. Here, where we are trying to form an idea of the outward appearance of the medieval Islamic city as discernible from the Geniza documents, suffice it to say that the constant sight of desolation presented by ruined houses cannot have annoyed the inhabitants excessively; otherwise, they would have found means to mend the situation. One reason for their attitude may be that houses were mainly, although by no means exclusively, oriented towards the inner court and not towards the street. Another reason may be a certain degree of otherworldliness which was not offended by such reminders of human decay and transitoriness.

The economic reason for the abundance of ruins in the medieval city was, to my mind, twofold: common partnership in houses and the neglect of their upkeep resulting from it; and the high cost of trained labor, which rendered repairs expensive while rents were comparatively low.

Housing, as reflected in the Geniza, displays a polarity characteristic of the society of that period in general, a society which was enterprising, mercantile and mobile, and, at the same time, tradition-bound and clannish. As a rule, various branches of an extended family lived in one house, and very frequently adjacent houses were in the possession of relatives or of one person. Marriages between cousins or other close relatives were concluded in order to reunite properties which had become divided through the process of inheritance. In short, housing was organized in such a way as to secure the coherence of the family in the wider sense of the word. On the other hand, a house was a monetary unit. Normally it was divided not physically but according to nominal shares. These shares frequently changed hands. Partnerships with outsiders, even members of another religion, were common. The right of preemption, intended to keep aliens out, often was waived or disregarded. The tension between these two opposing tendencies of mobility and family cohesion gave rise to the many contracts, lawsuits, and settlements which are recorded in the Geniza papers.

With very few exceptions, all documents coming from Egypt, whether issued by Muslim or by Jewish authorities, describe the houses concerned as being held in joint, undivided ownership. This means that the parts of a house, which normally formed the object of a contract, were units of account, not real segments of a building. A house was divided into twenty-four nominal shares, a division modeled on the twenty-four *qīrāṭs*, or parts of the dinar. The same division, as is well known, was also adopted in the apportioning of an inheritance in Islamic law. The shares transferred by sale or gift could be very small. In a carefully executed Arabic deed, a husband acquired for his wife 1/48 of a house from her brother. In the year of famine 1200 a woman sold 1/48 of a house for twelve dinars. Shares of 1/18, 1/16, 1/9, or 1/8 were commonplace. The majority of the transactions recorded concerned portions of a house amounting to 1/6 or more, which means that they normally were large enough to form separate apartments. In any case, we see that the institution of partnership, which in those days dominated commerce, finance, industry and employment, prevailed also in the ownership of houses.

When one or several partners in a house were absent for prolonged periods—for example, on a business trip to India or Spain—or were unable or unwilling to contribute to its maintenance, the house decayed and soon parts of it became uninhabitable. Repairs were expensive, as is proved by a great many accounts dealing with this subject preserved in the Geniza. A master mason, or a stucco-worker, or a layer of marble slabs or of clay pipes could receive five to six dirhems as his *daily* wages and, in addition, lunch worth one dirhem or more. However, for five dirhems *a month* one could rent a modest apartment, consisting perhaps only of one room and its appurtenances, as again proved by contracts of leases and other Geniza documents. It must be emphasized that in the prices of houses and amounts of rents paid, great differences can be observed in the relevant records. Still, with regard to many old houses the incentive to keep them in good repair must have been comparatively limited.

In addition to economic considerations, superstitious beliefs might have contributed to the abundance of ruins. A house in which a man was murdered or otherwise died a premature death, was regarded as ominous and was liable to be abandoned. To be sure, such beliefs are found in Islamic and Jewish literature, but have not been traced by me in the Geniza letters.

Living space must have been abundant. Vacant apartments are listed in the accounts of the pious foundations from the eleventh through the thirteenth centuries, as well as in documents referring to houses belonging to private persons. The government billetted its men in vacant rooms, in both private and communal houses, an unpleasant imposition which usually was warded off by a special payment. This state of affairs is illustrated by entries in the communal accounts such as: "Messengers to the government with regard to the house in the Tujīb quarter which is unoccupied" (1164), or: "Payment for the vacancy in the story which is held in partnership with the Judge—seventeen dirhem up to the end of the month" (1184). A scholar managing a house for friends writes that he cannot leave the place for this would cause them great losses if a vacancy occurred and the government billetted someone in it. The very fact that such a practice was rampant in both Fāṭimid and Ayyūbid times shows that empty apartments must have been common in those days. This is proved also by the tenor of letters referring to requests for renting a place which seem to assume that it was never difficult to find one. A letter from Jerusalem tells us that a whole suite was rented for a friend well in advance of his arrival in the Holy City, leaving it to him to choose the rooms in which he would prefer to stay. Our study of the amounts of rents in the Geniza reveals them as being comparatively low, which is best explained by the assumption that supply exceeded demand.

Besides houses, stores, and workshops, a great variety of other buildings are mentioned in the Geniza documents. There were many types of "houses" named after commodities such as apples, oil, rice, or gems in which wholesale business was conducted, as well as the caravanseries, colonnades, and bourses which largely served a similar purpose. As far as industry is concerned, besides the ubiquitous sugar factories, which were veritable landmarks, flourmills and oilpresses formed predominant features in the look of the city. The abattoirs were situated in the midst of lively business centers or residential quarters. Bathhouses were as common as theaters and movies in a modern town and partly fulfilled a social function not unlike them, inasmuch as they provided recreation in addition to hygienic facilities. Hospitals are very rarely mentioned and one has the impression that during the Fāṭimid period only one hospital was found within the orbit of the Geniza documents. Needless to say, synagogues are constantly referred to as places of worship, study, communal activities, lawcourts, and hospices, and the same applied, for the communities concerned, to churches and mosques.

It is astounding how rarely government buildings are mentioned in the Geniza documents. There were the local police stations and prisons, as well as the offices where one received the licenses occasionally needed, but even these are seldom referred to. The Mint and the Exchange are frequently referred to, but at least the latter was only semi-public in character, since the persons working there were not on the government payroll. Taxes were normally collected by tax farmers. Thus there was little direct contact between the government and the populace and consequently not much need for public buildings. The imperial palace and its barracks formed a city by itself, occasionally mentioned in Ayyūbid times, but almost never in the Fāṭimid period.

Government, although not conspicuous by many public buildings, was present in the city in many other ways. A city was governed by a military commander called *amīr*, who was assisted by the *wālī* or superintendent of the police. Smaller towns had only a *wālī* and no *amīr*. Very powerful, sometimes more powerful than the *amīr*, was the *qāḍī*, or judge, who had administrative duties in addition to his substantial judicial functions. The chief *qāḍī* often held other functions such as the control of the taxes or of a port, as we read with regard to Alexandria or Tyre. The city was divided into small administrative units called *rab'* (which is not the classical *rub'*, meaning quarter, but instead designates an area, or rather a compound). Each *rab'* had a superintendent called *ṣāḥib rab'* (pronounced rub'), very often referred to in the Geniza papers. In addition to regular and mounted police there were plain clothesmen, or secret service men, called *aṣḥāb al-khabar*, "informants" who formed a government agency independent even of the *qāḍī*, a state of affairs for which there seem to exist parallels in more modern times.

An ancient source tells us that the vizier al-Ma'mūn, mentioned above, instructed the two superintendents of the police of Fusṭāṭ and Cairo, respectively, to draw up exact lists of the inhabitants showing their occupations and other circumstances and to permit no one to move from one house to another without notification of the police. This is described as an extraordinary measure aimed at locating any would-be assassins who might have been sent to the Egyptian capital by the Bāṭiniyya, an Ismā'īlī group using murder as a political weapon. Such lists, probably with fewer details, no doubt were in regular use for the needs of taxation. In a letter from Sicily, either from its capital Palermo or from Mazara on its southwestern tip, the writer, an immigrant from Tunisia around 1063, informs his business friend in Egypt that he is going to buy a house and that he has already registered for the purpose in the *qānūn* (Greek *canon*) which must have designated an official list of inhabitants. With regard to non-Muslims, a differentiation was made between permanent residents and newcomers. Whether the same practice existed with respect to Muslims is not evident from the Geniza papers.

What were the dues that a town dweller had to pay to the government in his capacity as the inhabitant of a city, and what were the benefits that he derived from such payments? By right of conquest, the ground on which Fusṭāṭ stood belonged to the Muslims, that is, to the government (the same was the case in many other Islamic cities), and a ground rent, called *ḥikr*, had to be paid for each building. A great many deeds of sale, gift and rent refer to this imposition. For a small house in the Fortress of the Greeks, which sold for twenty dinars in 1124, a ground rent of one *qīrāṭ* had to be paid per month, which makes twelve *qīrāṭs*, or one-half dinar per year. Thus, the yearly ground rent represented 1/40 of the value of the property. Several accounts from the year 1183–1184, which register the revenue from rents of communal buildings and the *ḥikr* paid, show the latter being about 8 to 10 percent of the former. Should both series of payments refer to the same buildings, the ground rent here would be far smaller than the one paid for the small house in the Fortress of the Greeks. Perhaps there the houses had deteriorated while the ground rent remained the same. This would explain why, early in the thirteenth century, the community paid the ground rent for the poor living in that section of the city, who, being unable to pay it, presumably would have lost their domiciles. During the six Muslim months corresponding roughly to March through August 1218, the payment was 1005 *qīrāṭ* per month, i.e. about a thousand times as much as paid for the small house sold in 1124. In deeds of gifts it is repeatedly stated that the ground rent would be paid by the donor or by a third party, as when a husband gave to his wife a small house adjoining a larger one which was made a pious foundation and which bore the *ḥikr*, or when a father gave to his daughter an upper floor in his house on condition that she would receive the rents, in case she preferred not to live there, while the *ḥikr* would be borne by the father or his heirs. Pious foundation would let their ruins against the mere payment of the *ḥikr*. The expenses incurred by the tenant for the rebuilding would be recovered by him by living in the house or renting it to others. All this shows that the ground rent was an imposition very much felt by the population. Like other taxes it was farmed out.

Besides the ground rent, every month a *ḥarāsa*, or "due for protection," had to be paid to the government. The protection was partly in the hands of a police force, partly in those of the superintendents of the compounds, and partly was entrusted to nightwatchmen, usually referred to as *ṭawwāfūn*, literally, "those that make the round," but known also by other designations. As we learn expressly from a Geniza source, the nightwatchmen, like the regular police, were appointed by the government (and not by a municipality or local body which did not exist). The amounts of the *ḥarāsa* in the communal accounts cannot be related to the value of the properties for which they were paid, but it is evident that they were moderate.

In a responsum written around 1165, Rabbi Maimon, the father of Moses Maimonides, states that the markets of Fusṭāṭ used to remain open during the nights, in contrast of course to what the writer was accustomed from having lived in other Islamic cities. In Fusṭāṭ, too, this had not been always the case. In a description of the festival of Epiphany from the year 941 in which all parts of the population took part, it is mentioned as exceptional that the streets were not closed during that particular night.

Sanitation must have been another great concern of the government, for the items "removal of rubbish" (called "throwing out of dust") and "cleaning of pipes" appear with great regularity in the monthly accounts preserved in the Geniza. One gets the impression that these hygienic measures were not left to the discretion of each individual proprietor of a house. The clay tubes bringing water (for washing purposes) to a house and those connecting it with a cesspool constantly needed clearing, and there are also many references to their construction. The amounts paid for both operations were considerable. The Geniza has preserved an autograph note by Maimonides permitting a beadle to spend a certain sum on "throwing out of dust" (presumably from a synagogue). This may serve as an illustration for the fact that landlords may have found the payment of these dues not always easy.

In this context we may also draw attention to the new insights gained through the study of the documents from the Geniza about the social life of Cairo. Massignon had asserted, and he was followed by many, that the life-unit in the Islamic city was the professional corporation, the guilds of the merchants, artisans, and scholars which had professional, as well as social and religious functions. No one would deny that this was true to a large extent for the sixteenth through the nineteenth centuries. However, there is not a shred of evidence that this was true for the ninth through the thirteenth centuries. Others have come to the same negative conclusion through a critical scrutiny of Islamic literature. The novelty presented by the study of the Cairo Geniza documents consists in the simple fact that they demonstrate positively to us how artisans, merchants, and scholars actually lived and how their work was organized.

The term "guild" designates a medieval union of craftsmen or traders which supervised the work of its members in order to uphold standards, and made arrangements for the education of apprentices and their initiation into the union. The guild protected its members against competition, and in Christian countries was closely connected with religion.

Scrutinizing the records of the Cairo Geniza or the Muslim handbooks of market supervision contemporary with them, one looks in vain for an Arabic equivalent of the term "guild". There was no such word because there was no such institution. The supervision of the quality of the artisans' work was in the hands of the state police, which availed itself of the services of trustworthy and expert assistants.

Regarding apprenticeship and admission to a profession, no formalities and no rigid rules are to be discovered in our sources. Parents were expected to have their sons learn a craft and to pay for their instruction, and the Geniza has preserved several contracts to this effect.

The protection of the local industries from the competition of newcomers and outsiders is richly documented by the Geniza records, but nowhere do we hear about a professional corporation fulfilling this task. It was the Jewish local community, the central Jewish authorities, the state police, or influential notables, Muslim and Jewish, who were active in these matters.

As to the religious aspect of professional corporation, the associations of artisans and traders in imperial Rome, or at least a part of them, bore a religious character and were often connected with the local cult of the town from which the founders of an association had originated. Similarly, the Christian guilds of the late Middle Ages had their patron saints and special rites. The fourteenth century was the heyday of Muslim corporations, especially in Anatolia (the present day Turkey), which adopted the doctrines and ceremonies of Muslim mystic brotherhoods. One looks in vain for similar combinations of artisanship and religious cult in the period and the countries under discussion. On the other hand, we find partnerships of Muslims and Jews both in workshops and in mercantile undertakings, for free partnerships were the normal form of industrial cooperation, and were common as well in commercial ventures. The classical Islamic city was a free enterprise society, the very opposite of a community organized in rigid guilds and tight professional corporations.

Further, we have stated before that no formal citizenship existed.

The question is, however, how far did people feel a personal attachment to their native towns. "Homesickness," says Professor Gibb in his translation of the famous traveler Ibn Baṭṭūṭa "was hardly to be expected in a society so cosmopolitan as that of medieval Islam." Indeed the extent of travel and migration reflected in the Geniza is astounding. No less remarkable, however, is the frequency of expressions of longing for one's native city and the wish to return to it, as well as the fervor with which compatriots stuck together when they were abroad. On the other hand, I cannot find much of neighborhood factionalism or professional *esprit de corps,* both of which were so prominent in the later Middle Ages. Under an ever more oppressive military feudalism and government regimented economy, life became miserable and insecure, and people looked for protection and assistance in their immediate neighborhood. In an earlier period, in a free-enterprise, competitive society, there was no place for such factionalism. A man felt himself to be the son of a city which provided him with the security, the economic possibilities, and the spiritual amenities which he needed.

DISCUSSION

À propos of Professor Goitein's remarks about the ubiquitous presence of ruins in Fusṭāṭ, Professor Oppenheim observed that "in texts from the Old Babylonian period, especially from Sippar, ruins are very often mentioned. In fact, there were three different words in the texts for kinds of ruins. The presence of ruins is all the more strange in Sippar because the price of space was at a premium. Rent of houses was very high, and people lived in very narrow, cramped quarters. Also, ruins were almost always individually owned, characterized as the house of Mr. So-and-so, adjacent to the So-and-so ruin of Mr. So-and-so. I know only of two contracts from the Old Babylonian period in which the city as such sells a ruin which has no owner."

Professor Nader suggested, from her experience with Lebanese villages, that ruins may belong to people who have migrated. Professor Goitein agreed that this was true in old Cairo too: "There is the case of a man who went to India, and failed to pay his share in the expense of upkeep of a property held in partnership. Half, if not two thirds, of all references and contracts indicate that houses were owned in partnership." Professor Goitein saw this as the main source of the difficulty of the proper upkeep of houses. Professor Fernea confirmed the importance of this factor by his observations in both Iraq and

Nubia, where one finds, given "the Muslim pattern of inheritance and the segmentary nature of their kinship system, that a house is divided amongst its owners, especially if the mother's family happens to be involved and if she is a co-survivor with the siblings. Within a couple of generations, so many people are involved in the co-ownership of the household that disputes are frequent. They are unable to agree either upon a price for selling it or how to share the costs of repairs, and often it seems best to forget about it altogether and let the whole thing go to ruin."

The interpretation of the word "*Rīf*" was a matter of discussion in view of the conference's interest in the problem of urban and rural settings. Professor Goitein used the word to mean the Egyptian countryside in general, and spoke of the antagonism to the *Rīf* found in Jewish documents as an implication of urban antagonism to rural areas. Professor Lapidus, however, did not think the example conclusive because "the term *Rīf* in the usage of tenth century authors also has a more specific meaning. It refers to a *kūra*, an administrative district in Lower Egypt, in fact to the *kūra* of which Mahalla was the capital town. In this case *Rīf* could be a proper name. By the fourteenth century the word seems to apply to a major region in Lower Egypt, just as it applies now to the coastal zones of Morocco, as well as to countryside in general. In the context of the Geniza documents it may mean *province*—everything outside of Cairo." Professor Goitein added, "More exactly, everything outside Cairo, Fusṭāṭ and Alexandria."

Professor Brinner speculated that the objection to moving to the *Rīf* was a Jewish phenomenon: "What may have been *Rīf* to Jews might not have been *Rīf* to Muslims. Even a rather large town which did not have a considerable Jewish population and therefore did not have the appurtenances of a Jewish community would have been considered *Rīf* as far as the Jews were concerned. In a document where a woman stipulates that she would not go with her husband if he moved to the *Rīf* and mentions a specific community, it does not, to my mind, mean that Muslims would have the same attitude. Jews were unwilling to leave a city or a center where they could appropriately observe their religious restrictions." Professor Goitein replied that "Mahalla would meet the requirements of Jewish community life for it had a large Jewish community and several synagogues. I remember a letter written from Jerusalem which gives greetings to thirty-two persons there."

PART III

Contemporary Middle Eastern Cities

Introduction to Part III

Islamic society was by no means stagnant after the period of the eleventh to fifteenth centuries (reviewed in the previous sections) until the contemporary period, which is discussed below. Important changes took place in all aspects of Muslim life. New empires—such as the Ottoman, the Safavid, and the Moghul Empire in India, emerged, modifying ancient political institutions and developing new ones. Also, by the thirteenth or fourteenth century Islam had lost the religious and cultural vitality and originality which characterized it in earlier ages, but became instead more profoundly rooted in the lives of Middle Eastern peoples. By the sixteenth century, the impact of Europe's economic competition, with its profound effects on local trade and industry, was already being felt in some parts of the Muslim world.

Yet no changes were so far reaching as those which took place in the eighteenth and nineteenth centuries. The old centralized imperiums were undermined by local powers, though the Ottoman Empire managed to recreate itself in the nineteenth century along more modern and even western lines. By the nineteenth century, important changes were also taking place in religious and communal life. The whole fabric of Muslim life was being radically called into question by new nationalist conceptions of community, by new reformist or secular values, by the decline of religious and craft fraternities, and generally by pervasive cultural and economic influences emanating from Europe.

In his paper, Professor Issawi considers one crucial aspect of these nineteenth and twentieth century changes—the growth of the cities. He begins by noting the historical rhythms of city growth and decline in the Middle East, and considers in what ways these tendencies are continued or changed in the modern era. Professor Issawi points to the traditionally high degree of urbanization in the Middle East and to the historic alternation of emphasis in urban development between coastal regions and inland areas. He also examines the first impact of European commercial influence in the nineteenth century which favored the growth of port cities without changing the overall degree of urbanization. Since 1920, however, the proportion of population living in cities has increased dramatically and port cities have declined relative to inland capitals. Professor Issawi then analyzes two features of the new urbanization: "overurbanization" in relation to industrial development and patterns in the regional distribution of city populations. His survey of the basic economic and demographic factors in the region sets the context for the further analyses by Professors Gulick and Abu-Lughod.

Professor Gulick concentrates on the urban-rural relationship, bringing to the subject a wealth of information and results of investigations by sociologists and anthropologists which are beyond the prospects of the historians. This is not to say that the data is unambiguous and certain. The more deeply one probes, the more subtle the problems involved reveal themselves to be.

Essentially, Professor Gulick rejects the notion of a folk-urban dichotomy as a deeply rooted stereotype, in both European-American and Middle Eastern cultures. No sharp holistic distinctions may in fact be made. Professor Gulick examines a number of "trait-complexes," social and cultural indices, for the Arab Middle East, to determine how closely urban and rural populations resemble each other. Analyzing farming, kinship, factionalism, attitudes toward women, household size, religion, and various social values, Professor Gulick finds strong similarities and common themes running through both the urban and rural segments of the society. Other criteria may of course reveal sharp distinctions. Despite the general congruence of Professor Gulick's and Professor Lapidus' views, however, these findings, based on contemporary evidence, cannot be applied directly to past situations. Many contemporary factors not operative in ancient times, such as the new technology of communications, tend to reduce cultural and social differences.

Having reviewed the field of common village and city subcultures, Professor Gulick re-examines the conflicting stereotyped impressions and moral attitudes of Middle Easterners toward city and rural life. He aptly touches on the profound ambiguities involved and some of the reasons for which these often mistaken impressions are held.

Turning to dynamic forces influencing village and city subcultures, Professor Gulick discusses the influence of immigration from village to towns and its impact on city social life. He points in particular to the limitations on the full assimilation of rural migrants into so-called "city" ways, and to the deep ties between city and hinterland created by population movements. Professor Gulick concludes by noting the persistence of traditional cultural patterns and the radiation of city influences to surrounding hinterlands as factors affecting the dynamic distribution of "trait-complexes" in the contemporary Middle East.

Professor Abu-Lughod treats similar problems, but within the context of developments in one specific city—Cairo. Happily, it is the same city revealed in Professor Goitein's study of the Geniza documents, but Professor Abu-Lughod's sources are no longer the handwritten letters and contracts of the Middle Ages; rather, they are census statistics collected by modern bureaucracies. On the basis of her analysis of these figures, Prof. Abu-Lughod forms a typology of socio-economic "styles of life" in the Cairo population, discusses their distribution in space and considers the main tendencies for change. Cairo appears as a mosaic of subcities and subcultures.

Rejecting the urban-rural dichotomy, Prof. Abu-Lughod insists on a third element. In contemporary Cairo she finds people who by occupation and associated life styles may be characterized as rural; people engaged in older handicrafts or in other small shop employment to whom she refers as traditional Cairenes; and modern urbanites. In the last century the agricultural sector has shrunk dramatically, and the traditional one has narrowed but by no means has it become inconsequential. Cairo's "modern" population grows apace. Until recently, migration has been an important factor in the changing balance of Cairo's population. Professor Abu-Lughod analyzes the effects of migration on city society from the demographer's point of view. In opposition to Professor Gulick, she prefers to stress the aspects of absorption and assimilation of migrants rather than the persistance of rural cultural ways. Appraising the directions of change, Prof. Abu-Lughod foresees a merging of modern, traditional, and even rural heritages, a reciprocal assimilation to form a new society in the Cairo of the future.

CHARLES ISSAWI

Economic Change and Urbanization in the Middle East

Towns are among the most expensive of human artifacts, and among the most durable. It is therefore not surprising that, in a region with as old a history as the Middle East, some urban patterns have shown great persistence. Of these two may be noted here: over the last two thousand years, most of the Middle East has been highly urbanized; and the urban center of gravity has tended to swing to interior regions except when a powerful Western influence (Greco-Roman, from 300 B.C. to 600 A.D.; European, from 1800 to the 1920s) has pulled it to coastal areas.

The Middle East entered the Modern Age at the beginning of the nineteenth century, with two marked characteristics in its urban pattern: a large proportion of town dwellers and a concentration of population in inland towns. The scattered population estimates available for the period around 1800, although tentative, point to a high degree of urbanization. In Egypt almost 10 percent of the population lived in towns of over 10,000.[1] In geographical or greater Syria, Aleppo had a population variously estimated at 150,000 to 250,000, and Damascus, at about 100,000, while Hama, Homs, Jerusalem and Tripoli had 10,000 inhabitants or more.[2] Since the total population

[1] Cairo had a population of about 250,000 (see estimates by Volney, Jomard and others discussed by M. El-Darwish in *L'Egypte Contemporaine*, (March, 1929); other towns with 10,000 inhabitants or over were Asyut, Mahalla, Damietta, Rosetta, Alexandria and Tanta. The total population of Egypt is usually put at 2,500,000 to 3,000,000, but was probably nearer 3,000,000.

Throughout this paper available figures have been reproduced with little or no criticism of their sources, and have been used for the calculations of various ratios. This implies a confidence in the accuracy of these figures which is certainly not held by the writer, but the orders of magnitude involved and the general conclusions drawn are probably correct. It is also useful to have the scattered estimates in one place.

[2] J. Sauvaget, *Alep*, (Paris, 1941), p. 238, and H. A. R. Gibb and H. Bowen, *Islamic Society and the West*, vol. I, part I, (London, 1950), 281. In the seventeenth century Aleppo had about 14,000 hearths (*Encyclopaedia of Islam* [new ed.], s.v. Halab).

of Syria was probably below 1,500,000,[3] this implies an extremely high proportion of population living in towns, 20 percent or more. Figures for Iraq are even less reliable, partly because plagues, famines and floods took heavy tolls and urban population fluctuated sharply, thus greatly affecting the estimates made by European observers. The figures given by Olivier, Rousseau, and Buckingham indicate a population of 50,000 to 100,000 for Baghdad, and of 50,000, or a little less, for each of Mosul, Hilla and Basra. The total population of Iraq may be estimated to have been under 1,500,000, which again would imply a high degree of urbanization—perhaps as much as 15 percent. As for present-day Turkey—whose population in 1800 may have been around 10,000,000[4]—it contained such cities as Istanbul, with a population of 400,000 to 500,000 in the sixteenth century, 600,000 to 700,000 in the seventeenth, and somewhat more by the beginning of the nineteenth century; Izmir, with over 100,000; Bursa, with probably 50,000 or more; and smaller towns like Erzerum, Konia and Ankara.[5]

For Iran there are only a few rough estimates. There are many signs, however, that the population of several towns had been greatly reduced from their mid-seventeenth century level. Isfahan, whose population had been put by Chardin at 600,000 (he adds that it was "*aussi peuplée que Londres,*" which at that time had 400,000 to 500,000 inhabitants) had probably half that number by 1800; Tabriz, which had been almost wholly destroyed by the earthquake of 1721, had 30,000 to 50,000 inhabitants in the years 1810–1812; the population of Qazvin, Mashad and Shiraz had definitely declined; Tehran, which had just become the capital of Iran, had some 15,000; and the population of Yazd was put at the rather high figure of 100,000

[3] Charles Issawi, *The Economic History of the Middle East,* (Chicago, 1966), pp. 209, 220.

[4] The 1831 "census" put the number of adult males (*erkek*) in Anatolia at 2,384,000, implying a total population of about 10,000,000 to which should be added that of Istanbul—see Enver Ziya Karal, *Osmanli imparatorlugunda ilk nufus sayimi,* (Ankara, 1943), p. 215.

[5] O. L. Barkan, "Essai sur les données statistiques des registres Ottomans," *Journal of the Economic and Social History of the Orient,* I (1958); Robert Mantran, *Istanbul dans la seconde moitié du XVIIe siècle,* (Paris, 1962), pp. 44–47, and Bernard Lewis, *Istanbul and the Civilization of the Ottoman Empire,* (Norman, Oklahoma, 1963), p. 102. Figures of 600,000 to 900,000 are quoted for "Greater Istanbul" in the period 1815–1844. (Niyazi Berkes, *The Development of Secularism in Turkey,* [Montreal, 1964], p. 141). Izmir had a population of about 90,000 in 1650, about 30,000 in 1700, some 100,000 in 1715 and 130,000 in the 1830's (Berkes, p. 141) and Paul Masson, *Histoire du commerce français dans le Levant au XVIIe siècle,* (Paris, 1896), p. 416.

in 1810.[6] One gets the impression that the country was somewhat less urbanized than Egypt, Syria or Iraq, although in the absence of any reliable figures for total population[7] no firm judgment can be made.

The import of these figures may be brought out by comparing them with cities in Europe. In 1800, the percentage of total population living in towns of 100,000 or over has been put at 7 percent in England and Wales, 7 percent in the Netherlands, 2.7 percent in France, 1.6 percent in Russia and 1 percent in Germany.[8] The proportion living in towns of over 5,000 has been estimated at about 25 percent in England and Wales, under 10 percent in France, and distinctly less in the other countries, with the exception of the Netherlands.As for the United States, the 1790 census showed that only 3.3 percent of the population lived in towns of 8,000 or more, a figure that rose to 8.5 percent by 1840 and 16.1 percent by 1860. It was only with the advent of industrialization and steam transportation that most of Europe and America overtook the Middle East in the degree of urbanization.[9] It may be remarked, parenthetically, that earlier in its history the Middle East had contained still greater cities. It is not clear, however, if this implied a higher degree of urbanization since the total population was also much larger.[10]

[6] See Note 25 below and *Encyclopaedia Britannica,* eleventh edition, s.v. Isfahan, Tabriz, Tehran and Yezd. For the general decline in Iran from the end of the seventeenth century see L. Lockhart, *The Fall of the Safavid Dynasty,* (Cambridge, 1958), and N. V. Pigulevskaya *et al., Istoria Irana,* (Leningrad, 1958), chapters 8 and 9.

[7] Estimates for the 1850's to 1880's range from 5,000,000 to 10,000,000. The least unsatisfactory estimates are by a Russian scholar, Zolotarev, cited in L. A. Subotoinskii, *Pensiya: statistika-ekonomicheskii ochenk* [St. Petersburg, 1913] who gives a figure of 6,000,000 for 1888 and by Sir A. Houtum-Schindler who cites 7,654,000 in 1881, of whom 26 percent were urban. See *Encyclopaedia Britannica* (eleventh edition), s.v. Persia; George N. Curzon, *Persia and the Persian Question,* vol. II, (London, 1892), 492–3; and Eteocle Lorini, *La Persia Economica,* (Rome, 1900), p. 378.

[8] A. Bonne, *State and Economics in the Middle East,* (London, 1948), p. 224. By 1900 these figures had risen to 38, 22, 14, 9, and 16 percent, respectively.

[9] As for the whole world, it has been estimated that in 1800 less than 2.0 percent of its population lived in cities of 100,000 or more inhabitants; in 1850, 2.3 percent and in 1900, 5.5 percent. (Eric Lampard, "The History of Cities in the Economically Advanced Areas," *Economic Development and Cultural Change,* [January 1955]).

[10] The population of such Hellenistic cities as Alexandria, Antioch and Seleucia on the Tigris, at their peak, has been estimated at around 500,000 by Beloch, Heichelheim and other authors, and that of Constantinople has been put at not under 500,000 by Andréadès—in Norman Baynes and H. Moss (eds.), *Byzantium,* (Oxford, 1948), p. 53. But more recent estimates, based on fuller

Many factors help explain the high degree of urbanization during the late Arab and Ottoman periods despite a marked economic and cultural decline. They include the absence of a strong rural-based feudal system, prevailing rural insecurity, more favorable treatment by the government of townsmen than of peasants, and pilgrim and transit traffic.

The history of the Muslim Middle East shows few examples of strong, independent feudal lords living in country castles. The Mamlūk military leaders and landlords dwelt in Cairo, Damascus and other cities. This meant that the sums raised by them in rents and taxes, as well as those accruing to the monarch, flowed to the capital or other large cities. This greatly increased the purchasing power of the large urban markets—and correspondingly diminished that of the rural—inducing a concentration of craftsmen, merchants and others in the cities.

The insecurity of the countryside, which increased in the seventeenth and eighteenth centuries when bedouins extended their raids to the Egyptian Delta and the coasts of Syria and Palestine and order broke down in Anatolia and Iran, caused many farmers to flee to the cities and led to the wholesale depopulation of villages. Volney reports that whereas over 3,200 taxable villages in the Pashalik of Aleppo were recorded in the old Ottoman registers, "at present the collector can scarcely find four hundred."[11] At the same time an appreciable proportion of the inhabitants of most cities consisted of

archeological data, have tended to reduce these figures to about 200,000 for Alexandria and Constantinople, and 100,000 for Antioch. (Josiah C. Russell, "Late Ancient and Medieval Population," *Transactions of the American Philosophical Society*, XLVIII [1958], 68–92). Of course the size of urban populations fluctuated sharply, because of wars, epidemics, earthquakes and other disasters.

As for the Islamic period, A. al-Dūrī (*Encyclopaedia of Islam*, new edition, s.v. Baghdad) gives a figure of 1,500,000 for tenth century Baghdad, but archeological evidence makes one wonder whether in fact the population could have been much over 500,000. (Robert Adams, *Land Behind Baghdad*, [Chicago, 1965]).

Cairo was almost certainly smaller. In this connection, two estimates made at the beginning of the fourteenth century may be quoted: Simone Segoli's of "more than 300,000" and Simon Simeonis' statement, "Cairo is twice as large as Paris, and has four times the population," quoted in Gaston Wiet, *Cairo*, (Norman, Oklahoma, 1964), pp. 72–74; Paris had a population of 84,000 in 1292, which "perhaps reached 90,000 before 1348." (Russell, *Late Ancient and Medieval Population*, p. 107). See also other descriptions of Cairo quoted by Raphail Wahba in Morroe Berger (ed.), *The New Metropolis in the Arab World*, (New Delhi, 1963), pp. 25–30. In the same period Aleppo and Damascus may have had 100,000 inhabitants each and Tripoli 20,000. (Nicola Ziadeh, *Urban Life in Syria*, [Beirut, 1953], p. 97). Marco Polo gave a figure of 80,000 for Aden, probably an overestimate.

[11] See Issawi, *The Economic History of the Middle East*, p. 260.

farmers who cultivated adjacent lands, but who lived within the city walls for protection—a practice that has continued to this day in some places, for instance, in Damascus. Other peasants fled to the cities in times of famine believing, rightly, that the government would not let townsmen starve and would somehow or other secure provisions from the countryside, even if it meant rural famine. Again Volney provides interesting examples.[12] The same process seems to have occurred in Iran during the Second World War.

In the absence of commercial or industrial development—or an increase in agricultural production, and a rise in the surplus available for urban use—the towns could not have continued indefinitely to absorb this rural influx if high urban death rates, caused by poor sanitary conditions and raised sharply every few years by an outburst of plague, cholera or other epidemics, had not reduced city population.[13]

As for pilgrim and transit traffic, it is sufficient to recall the names of such holy cities as Jerusalem, Mecca, Medina, Karbala and Mashad, and to draw attention to the persistence of some trade through the Middle East even after the diversion of most trade between the Far East and Europe to the all-sea route around the Cape.

The second characteristic of Middle Eastern urbanization, inland location, may be dealt with much more briefly. The vast majority of large cities in Greco-Roman times were seaports: Athens, Corinth, Carthage, Syracuse, Ephesus, Alexandria, Antioch, Rhodes, Rome (which was accessible to quite large ships), Constantinople, and so on. So were most large Western cities in early modern times: Genoa, Venice, Naples, Lisbon, Antwerp, Amsterdam, London, Philadelphia, New York, St. Petersburg, and Copenhagen. But in the Islamic Middle East most of the great cities were inland: Damascus, Aleppo,

[12] Issawi, p. 216.

[13] Four examples, out of many, may be given. In May–August, 1669, an estimated 150,000 died of plague in Aleppo (A. C. Wood, *A History of the Levant Company*, [London, 1935], p. 246). In Baghdad, in April, 1831, the death toll was 40,000 to 50,000 (S. H. Longrigg, *Four Centuries of Modern Iraq*, [Oxford, 1925], p. 266). In Cairo, in 1834–35, "not less than 80,000" died of plague (E. W. Lane, *The Manners and Customs of the Modern Egyptians*, [London, 1944], p. 3). In Salonica, according to Rabbi Samuel de Medina, "In 1548 a great pestilence ravaged the city causing the death of about 7,000 Jews. Other plagues followed in the years 1552, 1554, 1561, 1564 and 1568." (Morris Goodblatt, *Jewish Life in Turkey in the XVIth Century*, [New York, 1952], pp. 22–23). Earlier outbreaks had been even more devastating, for example the Black Death in the fourteenth century. For comparable developments in Japan see Irene B. Taeuber, "Urbanization and Population Change in the Development of Modern Japan," *Economic Development and Cultural Change*, (October, 1960).

Baghdad, Mosul, Cairo, Rayy, Nishapur, Isfahan, Tabriz, Tehran, Konia. In this, perhaps, the Muslims reverted to the older traditions of the Babylonians, Egyptians, Aramaeans and Persians. Most of the larger ports inherited from the Romans shrank to insignificance (for examples, Antioch and Alexandria). The major exception was of course Constantinople, whose superb location and great defensibility were fully appreciated by the Ottoman sultans, and Tunis, which grew in North Africa, on the site formerly occupied by Carthage.

Part of the explanation of this phenomenon is no doubt to be sought in the origins of Islamic towns. Many of those founded by the Arabs were camps on the edge of the deserts in which the Arabs maneuvered so effectively—Basra, Kufa, Fusṭāṭ, Kairouan.[14] Others had already served for many centuries as "desert ports," handling caravan traffic, as did Damascus, Aleppo, Kerman, and Yazd. Others were royal cities—Samarra, Baghdad, Cairo, Meknes—and few Muslim rulers felt anything but aversion for the sea. But trade has its exigencies, and one cannot help surmising that great Mediterranean seaports would have developed, or revived, had it not been for the seapower first of the Byzantines and then of the Italians.[15] In the Middle Ages, the loss of control over the Mediterranean to the Franks led to a shift in the center of gravity to the interior. Thus in the fourteenth century Aleppo bypassed Antioch, which was falling in ruins,[16] and Cairo conducted much of its international business directly and not through Alexandria, which was vulnerable to raids from Cyprus, Rhodes and other Christian strongholds.[17] The passage of the centuries merely accentuated the decline of Alexandria, Saida, Beirut, Antioch and the other seaports, except for Salonica and Smyrna, which began to revive in the sixteenth century, thanks to their large Greek and Jewish populations.

Other striking characteristics of Muslim towns will not be discussed here, since they fall outside the scope of this paper: the complete absence of city states; the almost complete lack of municipal self-government and institutions; the coexistence of various religious communities, each in its own ghetto; the shutting off of each quarter

14 Xavier de Planhol, *The World of Islam,* (Ithaca, 1959), pp. 3–4.

15 See Archibald Lewis, *Naval Power and Trade in the Mediterranean,* (Princeton, 1951).

16 Sauvaget, *Alep,* p. 165. "Her husband's to Aleppo gone, master of the Tiger," *Macbeth.*

17 Subhi Y. Labib, *Handelsgeschichte Ägyptens im Spatmittelalter,* (Wiesbaden, 1965), chapter 9.

from the other parts by inner walls and gates;[18] and the concentric layout, with the "nobler" crafts and trades located immediately around the mosque and the "baser" ones on the outskirts.

EVOLUTION: 1800–1920's

In the course of the nineteenth century the various parts of the Middle East were drawn, to a greater or lesser extent, into the international network of trade and finance. This entailed the immigration of European businessmen and technicians, the investment of foreign capital, the development of mechanical transport, and the shift from a subsistence to a cash crop agriculture. The introduction of modern hygiene led to a sharp population growth, and foreign competition resulted in the ruin of the handicrafts.

All these trends had marked effects on the location, size and structure of Middle Eastern towns. Perhaps the simplest way to put it is that the economy began to be oriented outwards, toward the export of its primary products, that transport was developed accordingly, with railway lines or steamboat services (in Egypt, Iraq, and Iran) pointing to the coasts, and that the alignment of towns shifted in consequence. Certainly the outstanding feature of the urban history of this period is the growth of "heterogenetic" seaports[19]: Alexandria (population in 1927—573,000), Port Said (104,000), Suez (41,000); Beirut (population in 1932—161,000), Tripoli (51,000); Jaffa (population in 1931—55,000), Haifa (50,000), Tel Aviv (47,000); Basra (population in 1935—60,000); Aden (population in 1921—57,000), Jidda and Bahrain; Abadan (population in 1937—60,000), Khurramshahr (30,000); Izmir (population in 1910—about 250,000). In North Africa this development was even more striking: one has only to think of Casablanca, Rabat, Tangier, Kenitra (Port Lyautey), Safi, al-Jadida (Mazagan), Tetuan, and Agadir in Morocco; Algiers, Oran, Annaba (Bone), Bejaya (Bougie), and Skikda (Philippeville) in Algeria; Tunis, Sfax and Bizerta in Tunisia; and Tripoli and Benghazi in Libya.

The growth of these "heterogenetic" North African seaports was due to the immigration of hundreds of thousands of Frenchmen,

[18] For the striking parallelisms in China, see Wolfram Eberhard, "Data on the Structure of the Chinese City in the Pre-Industrial Period," *Economic Development and Cultural Change*, (April, 1956).

[19] See Robert Redfield and Milton B. Singer, "The Cultural Role of Cities," *Economic Development and Cultural Change*, (October, 1954). The authors give examples of "heterogenetic colonial" cities, including Jakarta, Manila, Bangkok, Singapore, Saigon, and Calcutta.

Spaniards and Italians who came to constitute either a majority or a very large minority of their inhabitants. In the Middle East, immigration played a smaller role, except in Palestine, where the Jewish influx began to assume significant proportions in the 1900's, and in the Arabian seaports, which absorbed large numbers of Indians, Indonesians, Somalis and others.[20] It is worth noting that, at their peak in 1907, foreign citizens constituted 25 percent of the population of Alexandria and 28 percent of that of Port Said, although they formed only 2 percent of the total Egyptian population.

However, the rapid growth of seaports did not entail a corresponding rise in the overall rate of urbanization. A careful study by Gabriel Baer shows the combined population of the twenty-three main towns of Egypt to have risen from about 400,000 in 1821 to 1,015,000 in 1882; 1,454,000 in 1897; and 1,596,000 in 1907; their share of total population increased from 9.5 percent to 12.8 percent, 15.0 percent and 14.3 percent, respectively.[21] In Syria, the figures given by Ruppin put the number living in towns of over 10,000, at the outbreak of the First World War, at 1,600,000, or 25 percent of the total population.[22] However, his estimates of the population of some towns seem to be inflated, and it is doubtful whether the proportion of urban dwellers did in fact increase in the course of the nineteenth century. For Iraq, the calculations made by M. S. Hasan show an unchanged urban proportion: 24 percent in 1867, 25 percent in 1890, 24 percent in 1905 and 25 percent in 1930.[23] Evidence for Turkey and Iran is too fragmentary for any firm conclusions to be drawn. *The Encyclopaedia Britannica* puts the number of Iranian towns of 10,000 or over in 1910 at forty-five, with a combined population of 1,700,000; only Tehran (280,000), Tabriz (200,000), and Isfahan (100,000) had 100,000 inhabitants or over.[24] The urban population may therefore have been over 15 percent of the total (perhaps 10,000,000), and there is no reason to believe that this ratio was appreciably higher than it had been at the beginning of the nineteenth century.[25] As for Turkey, the 1927

20 Aleppo also took in perhaps 50,000 Armenians after 1914.

21 Gabriel Baer, "Urbanization in Egypt, 1820–1907." (Paper presented at Conference on the Beginnings of Modernization in the Middle East, Center for Middle Eastern Studies, University of Chicago, 1966.)

22 A. Ruppin, *Syrien als Wirtschaftsgebiet,* (Berlin, 1916), pp. 187–88.

23 M. S. Hasan, "Growth and Structure of Iraq's Population, 1867–1947," *Bulletin of the Oxford Institute of Statistics,* (1958).

24 *The Encyclopaedia Britannica,* s.v. Persia.

25 The estimates quoted by Lord Curzon (*Persia*) would indicate a rise in the population of Tabriz from 30,000–50,000 in 1810–1812 to 170,000–200,000 in 1889 (Vol. I, 521) and that of Tehran from 120,000 in 1869 to 200,000 in 1889 (Vol.

census, which put the total population at 13,648,000, showed that there were five towns with over 50,000 each and with a combined population of 1,055,000: Istanbul, 691,000; Izmir, 154,000; Ankara, 75,000; Adana, 73,000; and Bursa, 62,000. There were another sixty towns with more than 10,000 inhabitants and with an aggregate population of 1,400,000.[26]

The very slow growth of urbanization in the Middle East at a time when the more advanced regions were rapidly increasing their town populations (see footnote 8) may be explained by two factors. In the first place, the growth of some seaports was partly achieved at the expense of other towns. Thus Cairo's trade was partly diverted to Alexandria and Port Said, as was that of Rosetta and Damietta.[27] Similarly Beirut took over much business formerly transacted in Damascus and Aleppo, as well as in such small ports as Saida. In North Africa there was also a decline in the relative importance of Fez, Meknes, Constantine and Kairouan. Second, the decline of the handicrafts—a phenomenon also observable in North Africa—certainly slowed down the growth of such towns as Aleppo, Damascus, Baghdad, Cairo, Tabriz, Isfahan, Bursa, Amasia, and Diyarbakir.[28] Sometimes it may even have caused an absolute decline: thus the Russian consul in Beirut, K. M. Bazili, states that the population of Aleppo dropped from 150,000 in 1820 to 80,000 in the 1840's, and that of Damascus from 120,000 to 80,000. However, although this decline has been attributed to the outflow of craftsmen from these cities,[29] it is not clear how much confidence can be placed in such estimates and whether the decrease was due to economic or other causes. These adverse effects were only partly offset by the favorable impact on urban growth of increasing centralization and bureaucratization, the rise in

I, 333), and a fall in that of Isfahan from 200,000–400,000 in 1784–1811 to 70,000–80,000 in 1889 (Vol. II, 43), and in that of Yazd from 100,000 at the beginning of the 19th century and 40,000 in 1860–1870 to 70,000 in 1889 (Vol. II, 240); that of Mashad is given at 45,000 for both 1830 and 1889 (Vol. I, 163). In 1796 Olivier had estimated the population of Tehran at less than 15,000. The population of Hamadan was put at 40,000 by Ker Porter in 1820 and at 20,000 by Curzon in 1889 (*Encyclopaedia of Islam* [new edition] s.v. Hamadhan). Kinneir stated that in 1813 Kermanshah contained 12,000 houses (Laurence Lockhart, *Famous Cities of Iran,* [Brentford, 1939], p. 55), and Morier put the population of Shiraz in 1811 at 19,000 (Arthur Arberry, *Shiraz,* [Norman, Oklahoma], p. 60).

26 In North Africa, however, where European immigration was large and flowed mainly to the cities, urban growth was far more rapid than that of the total population and the percentage of town-dwellers rose sharply.

27 Baer, "Urbanization in Egypt, 1820–1907."

28 Issawi, *Economic History of the Middle East,* pp. 41–59.

29 I. M. Smilianskaya, "Razloshenie feodalnikh otnoshenii v Sirii i Livane v seredine XIX v," *Peredneaziatskii Etnograficheskii Sbornik,* (Moscow, 1958).

national incomes, and the spread of Western-style living. And while the combined effect of all these factors was to slow down the rise in the numerator (the size of the urban population), that of the denominator (the size of the total population) was rising steadily. The expansion of agriculture, due to the establishment of law and order, the improvement of transport, the extension of the cultivated area, and the introduction of more valuable crops meant that employment opportunities in the countryside were growing sufficiently fast to absorb the increment in rural population. Migration to the towns was therefore rather small until the First World War.

One last point must be made. As Professors Abu-Lughod and Safran have observed, the composition of the population of the towns changed markedly during this period. To the extent that town-dwelling farmers gave way to persons engaged in more truly "urban" occupations, there was an increase in the degree of urbanization which is not revealed by the statistics.

It remains to add that these developments had a drastic impact on the traditional structure of Middle Eastern cities. New quarters in the Western style were erected alongside the old, throughfares were built through the old quarters,[30] the various communities began to emerge from their ghettoes and mingle, and tentative steps were taken towards municipal self-government.[31]

EVOLUTION SINCE THE 1920'S

The main trends that have affected economic development and urbanization in the Middle East during the last forty years are: the population growth, the discovery and exploitation of oil, the impact of the Second World War, the exodus of foreigners and minority groups from certain countries, the decline in the foreign trade of some countries relative to their gross national products, foreign aid, and, following independence, the use of state power to promote industrialization and other forms of economic development. The results have been a marked acceleration in urbanization and a reversal of the shift of population concentrations from inland cities to the coast. The ratio of urban (variously defined) to total population has risen at an accelerating rate. In Egypt it was 21 percent in 1917, 24 percent in 1937, 30 percent in 1947, 37 percent in 1960 and 40 percent

[30] Janet Abu-Lughod, "Tale of Two Cities: The Origins of Modern Cairo," *Comparative Studies in Society and History*, (The Hague, July, 1965), and M. Clerget, *Le Caire*, (Cairo, 1934).

[31] *Encyclopaedia of Islam*, s.v. Baladiyya.

in 1966. In Turkey it rose from 24 percent in 1927 to 25 percent in 1945, 32 percent in 1960 and 34 percent in 1965, according to official statistics; if the term urban is restricted to places of 10,000 inhabitants or more, the figures are 16 percent in 1927, 18 percent in 1945, and 25 percent in 1960. Comparable figures are not available for other countries, but the same phenomenon has undoubtedly occurred in Iran, where the population of Tehran rose from an estimated 300,000 before the First World War to 500,000 at the outbreak of the Second World War, 1,000,000 in 1950, 1,512,000 in the 1956 census, and about 2,700,000 in the 1966 census. Table I shows the degree of urbanization in the Middle Eastern countries in the 1950's. The reasons for this rapid growth in town population, which is almost certain to continue, are discussed below.

TABLE 1

Urbanization in Middle Eastern and North African Countries in the 1950's[a]

Country	Year	Percentage of Total Population		
		Urban	in cities of 20,000+	100,000+
Morocco	1952	...[b]	27	19.2
Algeria	1954	22.9	...[c]	9.9
Tunisia	1956	35.6	18.2	10.8
Libya	1957	...	...	15.0
Jordan	1952	37.7	...	8.1
Saudi Arabia	1954	...	...	8.4
Syria	1955	...	...	28.9
Sudan	1956	8.3	...	2.4
UAR	1957	35.8	36.9	22.4[d]
Iraq	1957	37.3	...	14.5
Kuwait	1957	...	...	50.6
Lebanon	1958	...	...	33.2
Israel	1954	56.4	52.0	36.4
Turkey	1955	25.0	...	10.0
Iran	1956	33.0	...	16.5

Source: Hubert Morsink, "al-Numūw al-ḥaḍarī. . . ." *Al-Abḥāth*, (Beirut), June, 1965; [figures on Iran, Israel and Turkey added].

a) Each entry covers the following ones in the row; thus in Tunisia 35.6 per cent of the population was urban; of these, 18.2 per cent lived in cities of 20,000 or more (including those of over 100,000) and 10.8 per cent in cities of 100,000 or more.

b) 29.3 in 1960

c) 14.1 in 1948

d) 1958

The notation ... means figure unavailable.

The relative decline in the importance of seaports compared to inland towns is due to many factors. In some, foreigners and minority groups formed a majority—or a very large minority—of the population and their exodus was bound to leave a large gap, for instance in Izmir, Istanbul, Alexandria and Port Said. In some countries, such as Egypt and Turkey, the failure of foreign trade to grow as fast as gross national product has helped to reduce the relative importance of seaports.[32] So has the phenomenal growth of capital cities, discussed below. The development of newly discovered mineral deposits and the growth of towns near them, e.g. Abadan, Kirkuk, and Aswan, and some deliberate attempts to decentralize industry have also contributed to move the urban center of gravity inland.

CHARACTERISTICS OF MIDDLE EASTERN URBANIZATION

Three characteristic features of present-day Middle Eastern urbanization are "overurbanization," the great size of the primate cities, and the emergence of megalopolitan centers.

There is no doubt that the Middle East is distinctly more urbanized than most other underdeveloped regions. The figures in Table I may be compared with the following table:[33]

TABLE 2
Percentage City Population, 1950

	in cities larger than	
	20,000	*100,000*
World	21	13
South America	26	18
Middle America	21	12
Asia (excl. USSR)	13	8
Africa	9	5

Perhaps more striking is the fact that in 1950 the Middle East, with a little over 3 percent of the world's population, contained four of the world's forty-nine million-plus cities; if Europe, Anglo-America and the U.S.S.R. are excluded, the number of such cities is reduced to twenty-nine. In other words, the Middle East contained

32 This is a widespread phenomenon in the underdeveloped countries and is well analyzed by Karl Deutsch and associates in "Population, Sovereignty and the Share of Foreign Trade," *Economic Development and Cultural Change,* (July, 1962).

33 Kingsley Davis and Hilda Hertz, "The World Distribution of Urbanization," *Bulletin of the International Statistical Institute,* No. 3, (1954).

14 percent of the large cities of the underdeveloped world.

The question has been approached from another angle—in some sense, urbanization in the Middle East has far outstripped industrialization.[34] The reasons for the very slow growth of industrialization until the last few years will not be considered here. Those for the acceleration of urbanization can be studied together with a closely connected phenomenon: the abnormally low share of the agricultural sector in gross national product, compared to other countries at the same level of per capita income, and—except in the oil-producing countries—the abnormally high share of the service sector.[35] First there is the rapid population growth—accelerating from 1 percent per annum in the 1920's to 2 percent in the 1940's and now approaching 3 percent. In several Middle Eastern countries—notably Egypt, Israel, Jordan, Lebanon, and Saudi Arabia, but also, to a lesser degree, Syria and Turkey—it is no longer possible to extend the cultivated area except by means of costly irrigation works. Hence the growth in the supply of agricultural labor has greatly exceeded the rise in demand for it, and millions have been pushed from the countryside to the towns. In some countries with no land shortage, such as Iran and Iraq, the land tenure system, by depriving the vast majority of peasants of land, induced the same exodus.

Second, the exploitation of the region's vast oil resources has given its governments huge revenues which at present exceed two billion dollars a year. A very large proportion of this revenue is spent in the cities, especially in the capital, thus making it possible to sustain a much larger urban population than would otherwise be feasible.

The traditional transit, tourist, and pilgrimage services performed by the Middle East, thanks to its location and history, have greatly increased in the age of pipelines and airplanes. Here too the benefits accrue to the urban centers: ports, airport cities, places of tourism or pilgrimage. And where such services yield substantial revenue to the goverment, as does the Suez Canal, for example, the process noted in the previous paragraph occurs.

34 For a review of the literature and the suggested criteria see N. V. Sovani, "The Analysis of Overurbanization," *Economic Development and Cultural Change,* (January, 1964). For Egypt, see Janet Abu-Lughod, "Urbanization in Egypt," *Economic Development and Cultural Change,* (April, 1965). The controversy seems to have been started by Kingsley Davis and Hilda Hertz, "Urbanization and the Development of Pre-Industrial Areas," *Economic Development and Cultural Change,* (October, 1954).

35 For a table and an interesting discussion see Frederic Shorter, "The Application of Development Hypotheses in Middle Eastern Studies," *Economic Development and Cultural Change,* (April, 1966).

Foreign aid—which the Middle East has taken in greater abundance per capita and from a wider variety of sources than any region in the world—has operated in the same direction.[36] The initial recipient of such aid is the central government, the bureaucracy in the capital city, and whether the aid is ultimately used for defense expenditure or economic and social development, the ultimate beneficiaries are mainly the urban population. Where the aid is extended in the form of food the connection is even more obvious, since foreign food makes it possible to support a larger number of city people without a corresponding increase in agricultural production.

The last factor is the course and nature of industrialization in the Middle East. In Europe and the United States during the "eotechnic" phase of technology—the era of wood, wind and water—industry grew up in the countryside, in search of water power; thus in Manchester, which had grown into an important textile manufacturing center, "as late as 1786 only one chimney rose above the town's roofs."[37] During the "paleotechnic" period—the era of steam and coal—industries had to concentrate in seaports, river ports or towns lying in lowlands, since coal and ores could reach them only by water or railroad. The "neotechnic" phase of electricity, internal combustion and chemistry, has once more made it possible to set up industries in rural areas, a trend that has been evident in the last few decades. In the Middle East, however, political, economic and social factors have led to the concentration of industries in cities.

In Europe and the U.S. it has been observed that the ratio of the size of cities follows a definite rule, the so-called rank-size rule. Thus in the U.S. the rank of a city (e.g. Chicago, 2; Los Angeles, 3; Philadelphia, 4) multiplied by its population gives a sum very close to that of the population of the largest city, New York.[38] More generally, the formula is $S_R = \frac{A}{R^n}$, where A is the size of the largest city, R is the rank of a given city and S_R is the size of the city of that rank. No satisfactory explanation has been given for this phenomenon. Since, however, it did not exist in the sixteenth to eighteenth centuries in

36 For comparative figures see H. B. Chenery and A. M. Strout, "Foreign Assistance and Economic Development," *American Economic Review*, (September, 1966).

37 E. Lampard, "History of Cities in Economically Advanced Areas," *Economic Development and Cultural Change*, (January, 1955).

38 See Rutledge Vining, "A Description of Certain Spatial Aspects of an Economic System," *Economic Development and Cultural Change*, (January, 1955), and literature cited therein. To what extent this pattern applies to France, where Paris continues to dominate the scene, is a question worthy of consideration.

Europe (where the population of London—Cobbett's "Great Wen"—Paris, Vienna and other giants were many times as large as that of the next biggest city), it may be surmised that it is connected with the diffusion of transport, industry, wealth and education over the whole area of an advanced country.[39] Thus in the industrialized countries, including the Soviet Union, indices of industrial concentration show a marked decline with the growth of total industrial production over time.[40] Similarly the spread of "provincial" universities indicates the diffusion of educational facilities in these countries. Such a diffusion may underlie the hexagonal and circular schemas of central cities described by Walter Christaller, which in turn may help to explain the existence of "a series of city tributary areas arrayed according to the rank-size rule."[41]

In the Middle East, however, the present-day pattern is much closer to that of the pre-industrial West. In most countries a giant capital city towers above the rest: Tehran, with a 1956 population of 1,512,000 compared to Tabriz, with 390,000, and Isfahan, with 255,000; Baghdad, with about 700,000 compared to Mosul, with under 200,000; Beirut, with about 400,000 compared to Tripoli, with a little over 100,000; and the Three Towns of Khartoum, Umdurman and Khartoum North with 240,000,[42] which are the only urban center of any significance in the Sudan.

In other countries two cities dominate the scene: Cairo, with 3,346,000, and Alexandria, with 1,513,000 compared to Ismailiya, with 276,000; and Damascus and Aleppo, with over 400,000 each compared to Hama, with about 170,000. It may be added that North Africa shows the same pattern, with Casablanca, Algiers, Tunis and Tripoli outclassing all the other towns in their respective countries. In only one country does the rank-size rule seem to hold at all—Turkey, where in 1965 the figures were: Istanbul, 1,751,000; Ankara, 902,000; Izmir, 417,000; Adana, 291,000; Bursa, 213,000; and Eskisehir, 174,000. Interestingly enough, the 1959 census shows that in Soviet Central Asia the rank-size rule holds quite well: Tashkent,

[39] For still earlier periods, however, the rank-size rule seems to hold somewhat better (Russell, *Late Ancient and Medieval Populations*, pp. 68–70).

[40] It is evident that the enormous growth in industrial production, education, and so forth, in the advanced countries had to be accompanied by diffusion since it would have been physically impossible to locate all the factories, colleges, and other institutions in a few centers.

[41] Edgar M. Hoover, "The Concept of a System of Cities," *Economic Development and Cultural Change*, (January, 1955).

[42] See Peter McLoughlin, "The Sudan's Three Towns," *Economic Development and Cultural Change*, (October, 1963; January, 1964; April, 1964).

911,000; Alma Ata, 455,000; Dushanbe, 224,000; Frunze, 217,000; Samarkand, 195,000; Ashkabad, 170,000; Chimkent, 153,000; Andijan, 129,000; Namangan, 122,000; Jambil, 113,000; and Khokand, 105,000. It also holds, somewhat more loosely, in Kazakhstan.[43]

The same concentration may be observed in all economic and social fields. Thus in 1958, 40 percent of the industrial establishments of the UAR were in "Greater Cairo" and 28 percent were in Alexandria; in Iran in 1963, 27 percent of industrial establishments, with 29 percent of employees, were in Tehran, and in 1965, 40 percent of electricity and 50 percent of oil products were consumed in that town and 50 percent of telephones were installed in it;[44] and in Iraq about 70 percent of factory employment was in Baghdad. Similarly, in 1957 Cairo had 48 percent of the UAR's installed electric power and Alexandria 15 percent, and for telephones the figures were 52 percent and 21 percent.[45] In 1953, Baghdad accounted for 65 percent of Iraq's physicians; in 1950, 25 percent of Turkish physicians were in Istanbul; in 1964, 33 percent of Iran's physicians were in Tehran; and in 1955, over 60 percent of the Sudan's technical personnel (physicians, engineers, etc.) were in the Three Towns. And it is only in the last few years that a beginning has been made in setting up institutions of higher education in cities other than the capital or the two main cities: Asyut; Izmir and Erzurum; Tabriz and Shiraz.

Here, too, the main explanation seems to be the administrative centralization and bureaucratization—and more recently the growth of economic planning—which has concentrated so much personnel and income in the capital city. In some countries the presence of foreign communities with high incomes has augmented the purchasing power of the one or two largest cities and enhanced their attractiveness. Industry has been drawn to these cities because of their purchasing power; their good communications; the relative abundance and reliability of their water supplies, electric power, repair shops and other external economies; and their access to government offices, in which so many decisions affecting economic activity are taken. It is only rarely that Middle Eastern governments have deliberately sought to decentralize industry, as was done by Ataturk in the 1930's, for political, social and strategic reasons, and as is being done in Iran today, in Isfahan, Shiraz, Tabriz, and other cities.

43 Edward Allworth (ed.), *Central Asia,* (New York, 1967), pp. 98, 107.
44 Bank Markazi Iran, *Annual Report,* (1966), and other sources.
45 Said El-Naggar in Berger, *The New Metropolis in the Arab World,* pp. 147–50.

As for education, in the Middle East the capital has always been the main cultural center (Baghdad, Cairo, Istanbul, Tunis, Fez). There are no examples of provincial Bolognas, Oxfords, Salamancas, Coimbras, Louvains, Leydens, Uppsalas, and Princetons, and few signs that this tradition is about to be broken.

It goes without saying that this centralization of talent and activity in the metropolis has unfortunate economic and social consequences. As regards economics, it seems likely that the benefits accruing from the "external economies" mentioned above are more than offset by the ever rising social marginal costs of the "social overheads" required by the expanding population of the primate cities—water, streets, sewers, schools, and so on; electricity is almost the only such service which operates under decreasing costs over a wide range of output. Socially, two aspects may be noted. First, the flight of talent to the metropolis is self-reinforcing, making it very difficult to persuade technicians, physicians, teachers and other *Kulturtraeger* to live in the provincial towns, much less the villages. Secondly, the "demonstration effect" of a given amount of modern culture concentrated in the metropolis is presumably much smaller than that of an equal amount divided among, say, half a dozen provincial towns, which could make an impact on a much wider surrounding rural population. Such considerations seem to have influenced Ataturk when he decided to set up many of his new factories—"Ataturk's minarets," as they came to be called—in the provincial towns rather than in Istanbul or Ankara.

Two further observations may be made. First, it may well be that the rank-size rule would apply on the regional scale, even if it does not apply on the national; the difficulty here is to identify the appropriate region (perhaps the Arab Middle East in which one can see an emerging pattern with the following order: Cairo, Alexandria, Baghdad, Beirut? Damascus? Aleppo? etc.). Second, one of the reasons for the concentration of city population in the primate cities may be Middle Eastern geography. The prevalence and pervasiveness of deserts has prevented the emergence of a broad continuum of cultivated, settled areas. Instead, a vegetation or population map of the Middle East shows archipelagoes of settlement surrounded by seas of desert. In these circumstances the hexagonal and other patterns noted in Europe are not likely to emerge, except perhaps after much greater development of transport.

The subject of the emergence of megalopolitan areas can only be touched upon. In the UAR the area between Alexandria and Cairo seems to be developing into one large megalopolis.[46] Another potential one is the coastal strip stretching from Gaza through Jaffa-Tel Aviv, Beirut, Tripoli and Latakia to Iskenderun or Mersin; at present, however, this coastal band is broken by four frontiers, all unfriendly and two closed. Last, the growth of Istanbul may extend over a large part of the Sea of Marmara, perhaps eventually joining Bursa. In this, as in other fields, the Middle East is erecting, on deep historical foundations, a building with many features of contemporary world civilization.

DISCUSSION

Professor Safran pointed out some methodological problems in comparing rates of urbanization over time or between different areas: "The study of comparative rates of urbanization raises some difficult questions. The most important problem may be put in the form of an abstract question: can we compare structural features without at the same time looking into functional aspects? Professor Issawi makes two kinds of comparisons: one between the rate of urbanization in the Middle East and in the West at a given time, and the second between the same area or the same city in the Middle East over time. With regard to the first type of comparison, it has been pointed out that Middle Eastern cities, especially before the nineteenth century, included a great deal of agricultural area, whereas in the West, because of the special privileges and legal condition of townsmen and the servile obligations of the rural population to their landlords, the urban pattern was not similar to the Middle Eastern one. Consequently, the size of population of Western cities involves an understatement, and the size of population of the Middle Eastern cities involves an overstatement. In making comparisons between the two, one confronts this problem.

"The second kind of comparison—involving the same city or area over time—presents a similar kind of problem. We know that contemporary cities include very little agricultural population in comparison with earlier times. Could not the apparent stagnation in the rate of urbanization, in what Professor Issawi calls the intermediate period, mask some shift in the occupational structure of city populations? Only when the transition to modern occupational patterns

46 Abu-Lughod, "Urbanization in Egypt," *Economic Development and Cultural Change*, (April, 1965).

was completed could urbanization in relation to the rural population become manifest.

Professor Abu-Lughod noted another factor which affected city growth: "When one talks about the growth of major cities in the Middle East today, one has the impression that migration is the chief source of this growth. This has traditionally been true. The city particularly before the nineteenth century lost population through mortality and therefore the deficit plus any growth, had to be made up from migration. This was still true at the beginning of the twentieth century. However, throughout the world, beginning around 1946, there was a "mortality revolution," in which death rates plummeted due to the use of antibiotics, DDT, and other technological changes which were supplemented by greater medical facilities. One of the interesting effects of this decline was that for the first time in many places, mortality in cities became less than in the country. In Cairo, for example, about fifty percent of the city's growth between 1937 and 1947 came not from migration, but from natural increase. By now about two thirds of every year's increase in population is due to an excess of births over deaths rather than to migration from the countryside. This is totally unprecedented."

Professor Lenczowski reviewed in more detail some of the political factors involved in city growth: "Professor Issawi points to the centripetal tendencies towards primate cities, and he indicates a number of factors which I would like to amplify. People flock to the cities because governments have money to disburse. Oil revenues and foreign aid produce a flow of funds from the government to the people, and of course those who are closer to the dispenser of money fare a little better.

"Another element is planning. Today every country in the Middle East, be it a patriarchial desert monarchy of the Saudi Arabian type, or a relatively advanced country like Egypt, engages in a good deal of economic planning, either under the socialist aegis or under the aegis of free enterprise. Government planning generates expenditures and stimulates the growth of bureaucracy, thereby increasing primate city populations. I made a study of the growth of the Libyan bureaucracy, which in the last four or five years has quadrupled in size, and now numbers (in a very small country) ten thousand officials, not to speak of all the servants in the ministries who are part of the patronage of quickly changing ministers. Here a very substantial part of the population has a stake in living in the capital city. Next are the contractors—native, foreign, and often alien Arab—who prey on the funds that the government is disbursing. Then foreign experts and officials grow in numbers by leaps and bounds. In Beirut there are an estimated 3,500 to 4,000 foreign nationals from other than Arab countries. These communities of government officials and experts from the U.N., from the World Health Organization, from the Food and Agriculture Organization, from UNESCO, and so on, have taken the place of the old foreign communities.

"Further, slum dwellers flocked to the cities because they hoped that they would find some employment and congregate in bidonvilles. For example, in Baghdad there is an accumulation of 60,000 to 100,000 *ṣarīfa*-dwellers.

"On the other hand, there are forces working to decentralize and reduce the importance of primate cities. One is the growth of oil cities, which of course follow the geography of oil discoveries. Some oil cities, because of location and/or company policy, integrate native and foreign populations, as has been done by imaginative housing schemes in Kirkuk. In contrast to this practice, Dhahran is divided into two communities, while some Persian Gulf towns are only partially integrated—such as Abadan.

"A second decentralizing force is that of deliberate planning, which, for example, is part of the five-year economic development plan of Libya. In Iran, too, with the conclusion of the steel plant agreement between the Iranian government and the Soviet Union in September, 1965, there is a deliberate policy of industrial decentralization which determined the construction of a metallurgical complex near Isfahan. Another example of decentralization is the creation of the Pahlavi University in Shiraz, as well as the universities in Tabriz and in Mashad.

"Another interesting experiment in this vein is being carried out in Israel—the cluster city idea. Israel has to face the economic and social problems of integrating immigrants from a variety of cultural backgrounds. The original idea was to mix them all together, but apparently this did not quite work. There were antagonisms between the European settlers and those from North Africa. As a solution the Israeli planners developed the idea of a cluster city arrangement—a central city comparable to a downtown service and administrative section surrounded by four or five cluster communities which we may call suburban. These communities were sometimes separated by a mile or two from the city, but remained part of the general complex. This decentralization may reinforce the tendencies noted by Professor Issawi toward the reduction of the relative importance of primate cities."

JOHN GULICK

Village and City: Cultural Continuities in Twentieth Century Middle Eastern Cultures

For at least 5,000 years most of the population of the Middle East has consisted of village farmers and urban gardeners, laborers, craftsmen, merchants, and institutional functionaries. Throughout this time, some cities have continued to develop, some have disappeared entirely, and others have been created in formerly uninhabited places. Urban populations have always been replenished by migrants from villages, while the political and commercial interests of city dwellers have continually been imposed on the villages. The influences between the rural and urban sectors have always been two-way processes.

Concomitant with this kind of history has been the development of a culture in which many details of life are essentially the same for villagers and urbanites. Although there are some very striking differences between them,[1] the village and city subcultures have many traits in common.

Those few social anthropologists, psychologists and sociologists who have seriously studied the contemporary Middle East seem to have accepted the idea that sharp, holistic distinctions between its rural and urban subcultures are impossible to make. They have put behind the polar folk-urban stereotype bequeathed by Robert Redfield and other representatives of the "Chicago School" of sociology. In so doing, they have counteracted the important strain of anti-urbanism among American social science intellectuals.[2] Whether all of them have entirely escaped the pro-urban biases of Middle Eastern intel-

[1] During recent centuries the village subculture of the Middle East has exhibited traits characteristic both of Eric Wolf's "corporate peasantry" and his "open peasantry" (E. Wolf, "Types of Latin American Peasantry: A Preliminary Discussion," *American Anthropologist*, LVII, 455). Simultaneously, the urban subculture could be characterized in terms of Gideon Sjoberg's "preindustrial city" (G. Sjoberg, *The Preindustrial City*, Glencoe, 1960).

[2] Morton and Lucia White, *The Intellectual Versus the City*, (Cambridge, 1962), pp. 1–2.

lectuals and Orientalists is, however, another question. To continue the advances made through this more realistic approach concepts like "urban" and "urbanization" should be defined in specific terms, rather than in abstract, global ones.

Thus, I shall outline some of the more important trait complexes of Middle Eastern culture, graded from most rural to most urban. Owing to scant data, the criteria are not rigorously quantitative, so their assigned positions are open to argument. They are presented in terms of what the anthropologist Max Gluckman calls "structural change"—the dynamic functions of apparently recurrent patterns of behavior. I shall then discuss some of the ways in which twentieth century Middle Eastern cities may be undergoing what Gluckman calls "radical change."

Before proceeding, certain factual differences between the Western and Middle Eastern rural and urban subcultures should be taken into account.

1) The rural settlement pattern of northwestern Europe and the United States, which was presupposed by the Chicago School, consisted of widely dispersed, often quite isolated, family farmsteads plus small, regional trading centers. Middle Eastern farmers, in strong contrast, live in villages which are tight clusters of houses and, in some cases, also small regional trading centers.

2) Middle Eastern farming grades off into pastoral nomadism, and the two ecologies interpenetrate each other as ways of life in many places.

3) Middle Eastern cities have fluctuated greatly in population during their histories; but on the whole they are small by Western standards. The kind of city that Middle Eastern theorists contrast with villages is much smaller than the kind of city which the Chicago School, for instance, contrasted with rural communities. As already indicated, Middle Eastern villages are quite different from American homesteads.

4) Middle Eastern cities lack large and heavy industries. Though commercially oriented, these cities have smaller and less complex bureaucracies than those created by Western heavy industry.

5) Many Middle Eastern cities, like many Western ones, are centers of governmental institutions. Middle Easterners' attitudes toward these institutions are changing, but they still contain the traditional opinion that the government is alien, corrupt, predatory, and repressive—not unlike the attitude of many Negroes toward government in the United States.

6) During the twentieth century Middle Eastern cities, like Western ones, have been growing at rapid rates. Nevertheless, the Middle Eastern population as a whole is still mostly rural. The consequences of rapid urban growth are not entirely the same as in the West. In part this is due to the relative lack of industrialization in the Middle East as compared to the West. In their contributions to this volume, Professor Abu-Lughod and Professor Issawi each stress the high urbanization of the Middle East It may be highly urbanized compared to preindustrial America and Europe in the past, but compared to present-day America and Europe it is not, as least as far as distribution of population is concerned.

RURAL-URBAN TRAIT COMPLEXES

Farming.

The primary village trait is the farming of a subsistence crop or a cash crop, but in either case the farmer is likely to be a tenant on land owned by a city-dweller. Farming may be supplemented by animal husbandry or a maintenance craft. Though typically considered rural, farming is also done on the fringes of cities by urbanites. For example, in 1956, 13 percent of the workers in the city of Kerman, Iran, were engaged in agriculture.[3]

Patrilineal Segmentary Kinship Structure.

In the Middle East, important and binding kinship ties are not limited to the members of individual households. Ideally, brothers are expected to support each other throughout life. There are stronger expectations of this sort among cousins and other relatives (especially patrilineal ones) than there are in the West. Although these expectations are complicated by a number of seriously conflicting emotions, they can have considerable force.[4]

Of course kinship ties do not have equal force for all Middle Easterners. Some people appear to be enmeshed in large kinship units in the form of tribes, with recognized leaders and collectively owned property. Others are members of large patrilineal name-groups which appear to have no other function than to provide an identity. Still others live in a network of kinship ties which do not appear to survive beyond the lifespans of a set of brothers. And there may well

[3] Paul English, *City and Village in Iran,* (Madison, 1966), p. 70.
[4] John Gulick, *Tripoli: A Modern Arab City,* (Cambridge, 1967), pp. 131–34.

be individuals whose effective kinship group does not extend much beyond the nuclear family. The concept of patrilineal descent, however, is firmly engrained in the culture and can potentially link anyone or any group to other individuals and groups. Political expediencies are frequently legitimized through the use of this principle.

In recognizing that there are variations among Middle Eastern people at any given time in regard to the size and extent of the kinship groups in which they are involved, I do not, however, acknowledge that there are consistent, predictable rural-urban differences in this matter. There is no substantial evidence in support of such an idea, but there are unsupported stereotypes. These allege that pastoral nomads are enmeshed in very large, widely ramified corporate tribes, and (at the other extreme) that urban people are only involved in nuclear families.

Contrary to these stereotypes, there are indications that all of the organizational variations of kinship mentioned above occur in Middle Eastern villages, and that probably all of them (including "tribal" groups) occur in Middle Eastern cities. Extended families—involving several separate households—are certainly operative in Middle Eastern cities. Lineages and clans, even if they serve only as name-groups or even if, as in Tripoli, Lebanon, they serve as the cores of major political factions, are also operative in cities. And al-Wardī, writing about Iraqi towns and cities of the recent past, refers to the "tribal" alliances between city dwellers and rural people.[5]

The idea that a prerequisite of enlarged kinship organization is the need for a co-operative labor pool—a vital necessity at certain times for farmers—coupled with the idea that urban industrial or office workers not only do not need such a large kinship organization but may actually find it a liability, has led to the notion of the logical "incompatibility" of industrialization and large scale kinship groups. These ideas may be appropriate for the rural and urban sectors of America in the nineteenth and twentieth centuries, but when applied to Southwest Asia, they require considerable modification. With the tightly nucleated Middle Eastern village, an extensive kinship organization is indeed helpful in providing a ready labor pool, but other emergency co-operative groups can also be organized among these people who are in such close communication with each other. In an urban culture where kinship loyalty and mutual support are clearly conceived virtues, large scale kinship organization is not necessarily a liability. To be sure, for the entrepreneur who wishes

[5] Ali al-Wardī, *A Study in the Society of Iraq*, (Baghdad, 1965), p. 181.

to invest his money in commerce and industry, the importunities of many needy relatives may be a liability. But, on the other hand, if a business or professional man has political ambitions (and many do), he can utilize his fellow clansmen as a core of constituents and supporters. Nepotism in government appointments (a virtue in the kinship-oriented value system) is one of the ways in which such a core is built and maintained. Mutual protection among kin-group members (against financial, medical, or social emergencies) is an important function which is still served by large kinship groups in Middle Eastern cities. Elsewhere I have emphasized that these group functions can remain important despite the presence of conflicting personal feelings among the kinsmen involved.[6]

Later in this paper, I mention the continuance of "tribal" organization and codes among the squatters in and around Baghdad. Hilali, the source of this information, expresses dismay that such behavior can occur in the city, although it should not be surprising considering the fact that the squatters are recent migrants from farming villages. His reaction probably indicates that there are differences among kinship structures in Iraq. But the significance of these differences in distinguishing urban from rural ways of life is quite another matter. Al-Wardī repeatedly refers to the presence of tribal structures in Iraqi towns and cities in the late Ottoman period.[7] In some cases, these appear to have been due to the presence of recent immigrants, though in other cases this is not clear. "Tribal," if it has a consistent meaning, means corporate kinship structure, and those Middle Easterners who are not participating in such groups always have a culturally available access to such kinship involvement. In Lebanon (often touted as the most Western and most "urban" of Arab countries) there has been a considerable movement over the past forty years for the legal incorporation of large, but hitherto unorganized, patrilineal kin groups into the social structure. In both cases the people involved seem to come from rural and urban areas. Although none of these newly incorporated lineages is likely to adopt the "tribal" system of indemnities in cash and women practiced among the *shurugiyya* of Baghdad, the same principles of group identity and organization are operative in both cases.

Factionalism.

The literature on Middle Eastern villages reveals the universality of factions—usually two opposed ones per village, although with great

[6] Gulick, *Tripoli*, pp. 135–37.
[7] al-Wardī, *passim*.

variations in intensity. While kinship groups play an important part in it, factionalism involves more than kinship loyalties and oppositions. Political, sectarian, and personal differences and conflicts are also very deeply involved. The kinship basis of factionalism is perhaps relatively more important in the villages, but sectarian and political prime-movers of disputes are not absent from villages. Nor, on the other hand, are kin-based factions absent from cities. Some factions are known to include *both* village and urban components. That there may also be an element of frustration–release (on any pretext) is a strong possibility, but it is clear that such frustration is not only generated in small villages.

Al-Wardī presents considerable information on factionalism in various Iraqi towns and cities from the late nineteenth century to recent decades. He emphasizes, first, the intense feeling of localism which is present in many city quarters, a feeling very similar to rural tribal loyalties which is frequently expressed in the form of aggressive street gangs.[8] He says that this sectionalism has tended to be less intense in Baghdad, owing to the effective presence of the central government. Even so, he reports battles which occurred between gangs of men and boys. For example, in 1912 there was an incident involving two of the quarters of the old city, Bāb al-Shaykh and Haydarkhāna, which arose out of differences over the Italian attack on Tripolitania. In 1920 there was an outbreak between some people from the quarter of Bāb al-Shaykh Banī Sa'īd and Karrada Sharqiyya. Other examples are adduced.[9] None of these feuds served to mobilize the whole city into two camps, but village factionalism does not necessarily do this, either.

Of particular interest are al-Wardī's descriptions of the factional problems involving the three major Shī'a Muslim shrine towns of Iraq: Karbala, Kazimiyya and Najaf. Kazimiyya (population 127,000 and now virtually a part of Baghdad but until recently a separate town) is the site of the double shrine of two of the twelve Imams. Karbala (population about 60,000) is the burial place of the martyr Husayn, and Najaf (population about 90,000) is the burial place not only of 'Alī, son-in-law of the Prophet, but also of countless Shī'as whose bodies were brought there from all over the Middle East. Sizeable foreign communities, mostly Iranian, are found in all three towns, but they are not involved in the cases of factionalism which al-Wardī describes.

In the eighteenth century, Najaf apparently mobilized itself for

8 al-Wardī, pp. 121, 178.
9 al-Wardī, pp. 183–84.

protection against Wahhābi attacks from the desert, but in the nineteenth century it was divided into two opposed named groups, a conflict which has lasted until recently. Karbala, too, particularly since the end of the Ottoman regime, has been similarly divided. Kazimiyya is apparently divided into several factions—at least, when Kazimiyya people make a pilgrimage to one of the other towns they go in several groups, each representing its own quarter.[10]

Early in the twentieth century a continuous feud began between people of Kazimiyya and Najaf. At one bloody incident in 1929, the Kazimiyya people were momentarily able to overcome their own internal differences. In 1945, efforts to negotiate an end to the feud failed, and al-Wardī says that it still goes on.[11]

Al-Wardī attributes these intra- and inter-city feuds to "tribalism." If "tribal" to him means rural, then the rural "impact" on these cities in considerable. However, I would prefer to view this material as evidence that factionalism runs equally through the rural and urban segments of Middle Eastern culture. Tripoli, Lebanon has two major socio-political factions which cannot be attributed to any rural influences, but which have roots in kinship group rivalries and in a long succession of political differences.

Physical aggressiveness on the part of males is an important element in Middle Eastern factionalism. Al-Wardī's descriptions of boys' gangs and adult gangsters in Baghdad have a flavor reminiscent of West Side "rumbles" and *machismo*, with a dash of Robin Hood thrown in. The occurrence of *al-lawt*, homosexual assault, in which the passive victim is shamed but the attacker gains prestige, may seem like a discordant element in this "heroic" picture.[12] It can, in any case, be linked with the sexual segregation of Middle Eastern culture, which is probably more intense than that which tends to accompany Latin Catholicism.

Another aspect of factionalism is sectarianism. Is consciousness of sectarian differences likely to be more intense in heterogeneous environments such as cities than in more homogeneous ones? Theoretically, it would seem reasonable to suppose so. In fact, however, there are Middle Eastern villages in which members of more than one sect actually reside. Such villages differ from cities in this matter only in the scale of their sectarian heterogeneity. The division of villages into two factions is reflected in the clustering of their mem-

10 al-Wardī, p. 191.
11 *Ibid.*
12 al-Wardī, pp. 324–25.

bers' residences. In cities, there are likely to be many (rather than two) residential clusters, neighborhoods or quarters which are culturally homogeneous, at least in reputation.

Sexual Alienation.

Males and females are considered to have radically different natures, the male being superior, but also vulnerable, to the female. Among the results of these attitudes are severe social segregation of the sexes and a very clear-cut sexual division of labor. Ironically, there are indications from studies of both village and urban subcultures that women often have more authority within their own households than they are supposed to have according to explicit norms. However, if there are strong patterns of affection and companionship between husband and wife, they must constitute a crypto-culture, for there are few external indications of their existence. Western ideas concerning sexual integration, equality, and affection are known to some Middle Easterners (especially Westernized urban ones), but they directly conflict with traditional attitudes and have had an operational effect only in certain limited sections of the population. Even now the huge cinema and restaurant crowds in Baghdad, for instance, are almost entirely male. And in Beirut, contrary to its reputation as a sophisticated international metropolis, the dating practices of women university students are restricted.

To the extent that non-sexual social relationships between men and women occur, they seem to be burdened by sexual jealousies coupled with alienation. Al-Wardī describes "home life" (he uses the English phrase in his Arabic text) in Baghdad as being sorely deficient.[13] While the men can and do spend much of their free time in coffeehouses with their male friends, the women's social outlets with their women friends are restricted primarily to their homes. Other sources on Iraq mention the *qabul* (women's eating, drinking, gambling, and gossiping session) which goes on for hours in private homes. Al-Wardī also says that the women may shoo their children out of the house, forcing them to play in the street, and thus to form street gangs. This ultimately adds new recruits to the existing factions. While there may be an element of caricature in his presentation, it cannot be lightly dismissed.

The observation that city women seem more often to be veiled than village or pastoral women has often been cited as evidence for more

[13] al-Wardī, p. 276.

severe segregation in the city. However, social segregation in the villages is clearly marked, and the participation of women in farming is due to dire necessity and is not an indication of any relaxation of the segregation principle. Not only are the pastoral tents divided into men's and women's parts, but the ideology of male superiority and female inferiority is expressed as vociferously by the pastoral people as it is by other Middle Easterners. Al-Wardī points out that it is the greater likelihood of encounters with strangers which prompts more veiling in the city, but he does not claim that the concept of segregation itself is any more or less strong there than in villages.[14]

Household size and domestic family structure.

There are two preconceptions about household size and family structure. One is that rural households are very large; therefore, rural families are very large, and are thus economically, emotionally, religiously, and socially very important, relative to other institutions. The other is that urban households are small; therefore urban families are small, and are thus not economically, emotionally, religiously, and socially important, relative to other institutions. In an examination of the actual structure of urban and rural households such presuppositions must be left behind. The average household size in the United States today is about 3.3 persons. The central tendency in the Middle East—judging from the miscellaneous sources which are available—is about six persons per household. One may think of the "typical American household" as consisting of a married couple and two or three children—four or five people per household, and, indeed, there are millions of such households. The average household size in the United States is, however, smaller than four or five people because there is a large percentage of one-person and two-person households. These consist of unmarried adults and of married couples living in domiciles without children. American culture does not discourage such people from setting up their own small households. Middle Eastern culture does. Unmarried young adults frequently continue to live with their parents or siblings; elderly couples or widows and widowers tend eventually to move into the households of married children of siblings. These differences in values and expectations probably account in good measure for the difference in household size between the United States and the Middle East. An additional contributory factor, however, is the greater number of children per married couple in the Middle East.

[14] al-Wardī, p. 279.

Studies of the internal structure of Middle Eastern households indicate quite clearly that the vast majority of them consist of nuclear families (parents and unmarried children) frequently augmented by other relatives who are also dependent upon the senior couple. Contrary to widespread belief, the statistically typical Middle Eastern household does not consist of the extended family (a patriarchal couple, their married sons, and sons' wives and children). Not only are such extended family households unusual in fact, but there are also indications that many Middle Easterners dislike the extended family as a household arrangement. Elsewhere, I have shown that an average household size of six persons in a given Middle Eastern community does not depend on the existence of extended family households.[15] There may be a cut-off point in the range of average household sizes above which the occurrence of a significant number of extended family households would be predictable, but I do not know where this point would fall, and the available Middle Eastern data offer no help. Furthermore, the available data do not permit us to conclude that village households are any larger than city households in the Middle East.

Let us consider the following figures from both rural and urban studies.[16]

By comparing rural and urban data from Cairo and Tunis, urban households are smaller than rural ones; but if one uses the figures for Baghdad and the Jordanian cities in the comparison, one could make the reverse argument. In addition to this inconclusiveness, there are various discrepancies, inconsistencies, and obscurities in these data.[17]

[15] John Gulick, *Social Structure and Cultural Change in a Lebanese Village*, (New York, 1955), pp. 48–49; Gulick, *Tripoli*, chapter 6.

[16] Figures in Table 1 are taken from sources given below: William Gorton et al., *Village Survey: Kasmie Rural Improvement Project*, (Beirut, 1953), p. 8; Charles Churchill, *The City of Beirut*, (Beirut, 1954), p. 3; Gulick, *Lebanese Village*, pp. 48–49; Louise Sweet, *Tell Toqaan: A Syrian Village*, (Michigan, 1960), p. 165; *Palestine Village Survey*, 1945, p. 432; Paul Stirling, *Turkish Village*, (New York, 1966), p. 39; Kahtan Madfai, "Baghdad," *The New Metropolis in the Arab World*, M. Berger, ed., (New Delhi, 1963), p. 51; A. Farrag, "Demographic Trends in the Arab World," *The New Metropolis in the Arab World*, (New Delhi, 1963), p. 17; Churchill, p. 30; Jane Hacker, *Modern Amman: A Social Study*, (Durham, 1960), p. 75; Gulick, *Tripoli*, p. 122; Paul Sebag, *La Hara de Tunis*, (Paris, 1959), p. 35; Janet Abu-Lughod and Ezz al-Din Attiya, *Cairo Fact Book*, (Cairo, 1963); Ahmad Sousa, *Atlas of Iraq*, (Baghdad, 1953), p. 9; Madfai, p. 51; Farrag, p. 14.

[17] Farrag's figures for Amman (for which he does not give a source, although the likely one is the 1952 Jordanian housing census) is 8.2 persons per household, clearly larger than Hacker's figures based on a sample survey in Amman. Yet source of information and size of the figures do not seem to be related. For example, Sebag's figures from Tunis, which are small, were based on a sample

TABLE 1

Rural and Urban Average Household Sizes

Rural Households			*Urban Households*		
location	*average size*	*source and year*	*location*	*average size*	*source and year*
8 Shī'a Muslim villages, southern Lebanon	6.60	Gorton, 1953	Beirut	5.76	Churchill, 1954
13 mostly Muslim villages, Biqā', Lebanon	5.50	Churchill, 1959	Amman: Christian	5.60	Hacker, 1960
			Muslim	7.00	Hacker, 1960
Eastern Orthodox village, central coastal Lebanon	6.00	Gulick, 1955	Tripoli (Lebanon): Christian	5.70	Gulick, 1967
			Muslim	6.70	Gulick, 1967
Sunnī Muslim village, northern Syria	5.80	Sweet, 1960	Tunis: Christian	4.80	Sebag, 1959
			Jewish	5.01	Sebag, 1959
			Muslim	4.80	Sebag, 1959

5 Palestinian villages	6.40	Palestine Village Survey, 1945	Cairo (1947)	4.70	Abu Lughod and Attiya, 1963
2 Turkish villages	5.80	Stirling, 1966	Baghdad (central city 1947)	9.20	Sousa, 1953
Iraq, national average (61% rural), 1957	5.65	Madfai, 1963	Baghdad area (1957)	6.16	Madfai, 1963
Sudan, national average (92% rural), 1955–56	5.00	Farrag, 1963	18 Jordanian cities	8.94	Farrag, 1963

But these studies are consistent in that they show Christian urban households to be smaller than Muslim urban households. This finding is also consistent with two other studies made in Lebanon which show the same results. They also show that there tends to be less difference in size between rural and urban Muslim lower class households than there is between rural and urban Christian middle class households.[18] Assuming that lower class Muslims in Lebanon are similar to lower class Muslims elsewhere in the Middle East, and then considering the fact that lower class Muslims constitute the majority of the people of the Middle East, this may further indicate minimal differences between villages and cities in regard to household size. In any case, the Muslim households of Amman and of Tripoli, averaging 7.0 and 6.7 persons respectively, are essentially the same in size as the village Muslim households which were listed above.

It is assumed that Middle Eastern households consist almost entirely of kinsmen, usually in some form of the nuclear family. However, observations have been made of urban slum households that

---

survey, but the figures for Cairo, which are equally small, were based on the complete, official census of 1947. Sousa's figures for central Baghdad, which are very large, were also based on an official and complete census, however, and not a sample survey. I do not know how to account for the difference between Sousa's figure and Madfai's figure on Baghdad. Sousa's does seem very large, yet I have checked through its components (average household sizes for every section of the city), and reasonable explanations can be given for them. On the other hand, though Madfai's figure of 6.16 persons per household is more "in line" with other Middle Eastern cities, he does not acknowledge his source for it. One reason for the difference in the two figures may be that while Sousa's definitely does not include squatter households, Madfai's probably does. There are some indications that the numerous squatter households are small, about five persons on the average.

One specialist on Egypt has told me that in his opinion the figure of 4.7 people per household in Cairo is not very different from rural Egyptian averages. If this is true, then the average Egyptian village household may be smaller than that of some other Middle Eastern countries. Another specialist on Egypt has suggested to me that 4.7 may itself be too low for Cairo owing to the possibility that the census enumerators overestimated the number of really separate households in the crowded, multistoried tenements of Cairo. Conversely, the typically large Baghdad house, with a courtyard (if old) or a garden (if new), and often consisting of two stories with only one definite entrance, may frequently have been counted as one household by census enumerators when in fact it sheltered more than one *ménage*—thus resulting in the large average of 9.2 persons per household in the 1947 census.

Gulick's and Hacker's figures on Tripoli and Amman are based on sample surveys, and Gulick's sample was very small (thirty-three cases).

[18] Terry Prothro, *Child Rearing in the Lebanon,* (Cambridge, 1961), p. 45; David Yaukey, *Fertility Differences in a Modernizing Country,* (Princeton, 1961), p. 169.

consist of non-relatives who rent space.[19] How numerous such households may be in the Middle East is presently unknown.

Unless more conclusive data are brought to light, the assumption that village households are consistently larger than urban ones is unwarranted. We can, however, accept the idea that both village and city households in the Middle East are significantly larger than North American ones. Lastly, while there is evidence of considerable variation in household size in the Middle East, this variation is not always expressed in terms of clear-cut rural vs. urban types. This should alert us to the probable existence of important similarities between rural and urban patterns of family behavior.

Individual Orientation Toward Social Relationships.

Because most Middle Eastern villages are small, every resident knows everyone else personally. To a considerable extent city dwellers (both migrants and native urbanites) tend to live in similar, personalized environments. Nepotism in government and business is an aspect of this kind of interaction, as are the highly elaborated courtesies of *wasta*—the use of intermediaries in adjudicating disputes or in seeking assistance from power figures.[20] Also, there are Middle Eastern traditions emphasizing the direct accessibility of power figures to the most humble visitor. Though often romanticized and mistaken for egalitarianism (which it most certainly is not), such accessibility is not imaginary. I know from my own experience that the office of the mayor of Tripoli, Lebanon, has no outer suite and no visible secretarial staff at all. And in 1965, when I visited a cabinet member of Iraq in his Baghdad ministry, I discovered that he had no more buffers between himself and the outside world than does an average department chairman in an American university. True, persons who occupy such positions in the Middle East are well versed in the arts of evasion; nevertheless, the opportunity for personal contact is there to a degree that is not found in comparable situations in the Western urban environment.

Although the Middle Eastern city dweller encounters more strangers than the villager, most of his activities are conducted in an individualized and personal style. In this respect village and city life are similar. Of course this style does not mean that the people con-

19 Shawqi al-Sibai, "Slum Family Life: A Study of Three Dwellings in the City of Port Said," (American University in Cairo, 1965), Unpublished M.A. thesis in anthropology.

20 al-Wardī, pp. 358–60.

cerned do not exploit each other, or use each other as objects, or hate each other bitterly.

This brings us to an apparent contrast between Middle Eastern culture (rural and urban) on the one hand, and contemporary Western culture on the other. Impersonality and anonymity are said to be among the evils of Western culture. Although there are also many "personal" relationships in Western urban life, they are overshadowed by two major problems of which "impersonality" and "anonymity" are, perhaps, only short hand labels. The first of these problems is commonly called cultural "pluralism." It consists of the sense of alienation and perplexity which many people feel as they observe others behaving in ways which they consider wrong. Not only do these others seem to behave improperly, but they sometimes appear to be rewarded rather than punished for their conduct. Their behavior, furthermore, is apparently not based on any identifiable group standards. Awareness of pluralism is greatly increased (if not exploited) by the mass media of communication. Similarly affected by the mass media is the second problem for which "anonymity" may be only a superficial label. This problem is a sense of individual helplessness in the conduct of large-scale affairs. This feeling occurs in spite of the stated values of Western culture which proclaim the individual's worth.

Traditionally, Middle Eastern urban and rural culture has had ways of coping with problems similar to these. Differing ethical standards are identified with different socio-cultural groups, to only one of which each individual can belong. His group is "right," and all others are "wrong." The sense of helplessness in larger issues is thus countered by personal contacts. Beyond this, however, the individual's social awareness has been limited to his immediate personal concerns, and his fate in general has been considered to be in the hands of God.

In current parlance, the first of these coping patterns consists of unbridled "ethnicity," and the second reflects complete lack of what Daniel Lerner has, in this very context of Middle Eastern culture, called "empathy."[21] Can a people increase their "empathy" and at the same time not have their "ethnicities" suffer? And if their "ethnicities" *are* undermined, how can they escape all the alienations inherent in "pluralism?"

One thing appears to be reasonably certain: if, as Lerner says,

[21] Daniel Lerner, *The Passing of Traditional Society: Modernizing the Middle East,* (Glencoe, 1958), pp. 49–54 and *passim.*

"empathy" may be a consequence of extensive exposure to the mass media of communication, then Middle Eastern villagers, just as much as city dwellers, may in the future become more "empathetic," since it is in the very nature of those media to transcend the limits of physical locality. So this may become another instance in which village and city dwellers of the Middle East share the same culture.

Commercialism.

Trading and marketing have always linked villages and cities. However, concentrations of diversified crafts are hallmarks of the urban scene, not the rural, although the differences may only be of degree.[22] Economic power certainly has been centered in the cities, and it continues to be as traditional economic patterns are altered by modern, Western-style financing, manufacturing, and distribution procedures.

The *sūq* is one of the most prominent characteristics of the traditional Middle Eastern city, and it has by no means disappeared from the modern scene. A *sūq* consists of a concentration of small shops along one or several streets which specialize in the same commodity but offer the customers different qualities at different prices. In many instances, the shopkeeper is not merely a retail salesman but is also a craftsman, as in the case of the shoemaker and the jeweler. Given enough stability, the system lends itself to the development of personal shopkeeper-client relationships lasting over long periods.

The presence of many such *sūqs* is what gives the Middle Eastern city much of its uniqueness and sets it apart from the village. Most cities, in fact, serve village hinterlands whose inhabitants must come to the city for certain products.[23] Apparently, this has led to a reputation for cheating and dishonesty among urban merchants. Al-Wardī states flatly that the influence of the profit motive in cities is what makes city people dishonest and changes villagers into dishonest people when they become city dwellers.[24]

But shops and shopkeeper-craftsmen are also present in villages, though in less variety and smaller number than in the cities. There is no evidence that the village shopkeeper is more honest in his dealings with fellow-villagers whom he knows personally than the city shopkeeper is in his dealings with members of his regular clientele. Neither is there evidence that the village shopkeeper is less likely to

22 English, *Iran*, p. 71.
23 English, chapter 4, *passim*.
24 al-Wardī, pp. 206 ff.

cheat a stranger than is an urban shopkeeper. Of course urban shopkeepers have more opportunity to cheat strangers than do village shopkeepers, but the fact that they have regular clienteles indicates that relative honesty and trust are also present in the city. In recognizing certain differences between rural and urban commerce, I question whether there are also significant differences in "commercial-mindedness" and its various ethical concomitants.

This discussion has not yet taken account of the great growth in Middle Eastern cities of Western style businesses and professions—banks, insurance companies, all of the businesses connected with modern means of transportation, hospitals, and so on. Their physical growth has resulted in new sections of Middle Eastern cities which resemble the American "central business district." Although such Westernized business districts in Tripoli and Baghdad, for instance, are also residential districts, they depart from the older style of shopfronts combined with houses—a characteristic of the traditional city which does not contrast radically with village styles.

While Westernization in Middle Eastern cities increases the contrast between them and villages, it does not necesarily have this effect. For example, the recent growth pattern of Middle Eastern cities has consisted of building steadily out from the center, with residences and retail commercial outlets in the same buildings—all of which perpetuates the older patterns. Among the largest of the Middle Eastern cities there are some exceptions. Cairo, for example, has some true suburbs in the Western industrial sense, and Baghdad has something resembling tract housing (but without supermarkets or shopping centers). For the most part, however, those aspects of Western urbanism which are ingredients of the megalopolis are absent from the Middle East.

In general, cities offer more facilities for diversion outside of the home than do villages. For this reason, the city tends, in Judaeo-Christian-Islamic terms, to be visualized as being "sinful" in a way that the village is not. Once more, we are reminded of the rural-urban moralistic stereotype. The clear indications that most city dwellers do not usually engage in orgies of food and sex any more than villagers do, appear to have had little effect on the stereotype.

Organized prostitution may be the most ancient of these urban facilities. Houses of prostitution have usually been concentrated, like most other commercial specialties, in the *sūqs* of the cities. Recently reformist governments have attempted to eliminate them. For example, in 1952–1953 the Iraqi government demolished the red light district of Baghdad which was in the old Maydan quarter. Displaced by a parking lot and bus stops, the prostitutes dispersed into various newer residential areas of the city. Demolition also currently threatens the red light section of Beirut, which, along with bus stations and auto repair shops, is concentrated in the downtown business district. A recent study of the prostitutes in Beirut indicates that they include more non-Lebanese persons and fewer people of rural origin than does a sample of Beirut housewives which was studied a few years earlier:[25]

	NON-LEBANESE *percent*	RURAL IN ORIGIN *percent*	SOURCE
Prostitutes	44.6	13.8	Khalaf, 1965.
Housewives	31.3	24.2	Churchill, 1954.

Khalaf discovered that half of the Beirut prostitutes began their careers as unwed mothers[26]—indeed they may have been forced into prostitution because of this, for being an unwed mother in the Middle East is a more difficult problem than in the United States. It is possible that there are more opportunities for a city girl to become an unwed mother than a village girl.

Belly dancers are an old tradition in the Middle East, but apparently until recently their performances were largely restricted to private parties. Al-Wardī says that the first belly dancer who appeared in a public commercial establishment in Baghdad did so in 1908.[27] Cabarets of Middle Eastern type (all male clientele) were just beginning at that time, at least in Baghdad. The development of Western style night clubs has been more recent, and caters to an equally restricted clientele consisting of foreigners and Middle Easterners. These are certainly urban institutions, but they have little significance for the culture as a whole. The same can be said of most restaurants and, to some degree, of hotels.

The cinema has made a great impact on city life in the Middle East, and it is indubitably one of the city's attractions for the villager. About one-third of a sample of industrial workers in Alexandria went to the movies at least once a week, and the indications are that more would go if they could afford it. Workers born in the city at-

25 Samir Khalaf, *Prostitution in a Changing Society: A Sociological Survey of Legal Prostitution in Beirut,* (Beirut, 1958), pp. 17, 31; Churchill, *The City of Beirut,* p. 47.

26 Khalaf, p. 23.

27 al-Wardī, p. 330.

tended more often than workers who had migrated from villages. This, however, does not necessarily indicate anything about differences in taste, for the migrant workers were less well off financially. Possible differences in taste between the migrant and city-bred workers were reflected by the greater preference of the former for leisure time activities in religious organizations and of the latter, in sports, but these preferences involved fewer than one-tenth of the sample. For the most part, both categories of workers preferred to stay at home or to go to a coffeehouse during their time off.[28]

Religious Behavior and Sectarian Institutions.

Theoretical discussions of normative Middle Eastern religious behavior patterns are rife in the literature. However, descriptions and discussions of actual practices (apart from weddings and funerals) are rather few. Nothing in this literature leads us to believe that Middle Eastern villagers and city dwellers differ essentially in the religious aspect of life, with the exception of the institutional features of Islam which are urban rather than rural. Other than this, there are no indications of significant differences in such matters as prayers, rites of passage and of intensification, belief in saints, the jinn—or the evil eye—and so forth. There are most assuredly no strong indications that Middle Eastern villages, as opposed to cities, can be characterized as "sacred" rather than "secular" communities.

While few systematic differences in religious behavior between village and city have been clearly demonstrated, sectarian institutions are almost entirely urban. In the traditional culture of the Middle East (before the beginning of Westernization in the nineteenth century) the cities were the centers of religious institutions to a greater or lesser degree, depending on their size. Here were all the greatest mosques of Islam, some of them shrines of such importance that they were the *raison d'être* for the cities surrounding them. Here, too, were all the libraries and institutions of higher learning. Al-Wardī feels that the preachers and theologians attached to these institutions contributed to urban life an element of "doubleness" (that is, fine-sounding ethical theories not realized in practice) which village and pastoral people were spared.[29] It is possible that what Al-Wardī has in mind is not so much the idea that city dwellers generally practice less of what they preach than do village-dwellers, but that among

[28] Hassan el-Saaty and Gordon Hirabayashi, *Industrialization in Alexandria*, (Cairo, 1959), pp. 118–23.

[29] al-Wardī, p. 293.

elite male groups in Baghdad, for example, it has been common practice to hold *majālis*, or *salons*, which purport to be theological discussions but also serve to facilitate making business and political deals.[30] Al-Wardī's criticism notwithstanding, the presence of these institutions in the city has conveyed a sense of glamor, prestige, and an image of sacredness not to be found in many villages.

Westernization has involved a relative de-emphasis of theologically oriented schools and a tremendous expansion of secular schools. The largest, most advanced, and most specialized of these are to be found in the cities, just as the older types were. Westernization does not appear to have altered the great importance of the shrine cities.

Elite Social Classes.

Middle Eastern culture, like most Asian agrarian cultures, is highly stratified, and repressive power tends to be held and exercised by small, elite classes. Mobility into and out of the elite is considerable, but the existence of the elite is a constant factor. Although the elite live in the cities, they may have strong village or pastoral ties, and their power as landlords is continually felt in the rural areas.

A Middle Eastern folk-urban stereotype, primarily based on the role of the urban elite, colors much indigenous scholarship. This double dichotomy characterizes urban life as civilized but degenerate, in contrast to the savage but noble nomadic culture, and relegates the great majority (villagers) to a status neither noble nor civilized. Perhaps there is a connection between the association of elite social classes with the city and the denigration of manual labor which is a feature of Middle Eastern values. Although much manual labor takes place in an urban context, peasant activities seem to receive most of the scorn, and it is obvious that the best place in which to escape peasant life is the city. Part of the denigration of peasant life is doubtless due to the effects of centuries of debt-ridden sharecropping. Another factor is the peasant's illiteracy, whereas in the cities literacy is necessary to gain power.

In short, the city is thought of as a place of prestige, regardless of the fact that actually most city-dwellers are poor and debt-encumbered.[31] In a very recent study, it has been found that members of the Arab minority in Acre, Israel, despite the many economic, emotional, and political disadvantages which they suffer there, prefer

30 Ibrahim Barudi, *Al-Baghdadiyyūn,* (Baghdad, 1958), *passim.*
31 Churchill, *The City of Beirut,* pp. 64–65.

to remain in the city rather than work in villages where there might be good opportunities for them.[32]

It is ironic, therefore, that Middle Eastern Muslim writers and philosophers have extolled the supposed virtues of the pastoral nomads, while the supposed vices of the city dwellers have been condemned. The city dwellers have, it would seem, no virtues except those which they may retain (in some mysterious fashion) from their supposed nomadic ancestors, while those nomads who move to the city are invariably corrupted. The city is positively associated with "civilization," and yet it is also unwholesome and effete, and survives only by means of constant infusions of new blood from rural areas, particularly the deserts.

This point of view was long ago expounded by Ibn Khaldūn and it is not surprising that his ideas tend to be greatly appreciated by the rural-urban dichotomists of the West.[33] We do not have to rely solely on old sources for such expressions. They are reported at considerable length in al-Wardī's recent book, despite the evidence of the following facts: In 1957, the population of Iraq was estimated to consist of 68 percent villagers, 30 percent city dwellers, and 2 percent pastoral nomads.[34] These proportions are generally the same as those claimed for the Middle East as a whole. They indicate a predominantly non-urban population, only a very minor segment of which, however, consists of pastoral nomads. There is some evidence that during the past century the pastoralists in Iraq, for example, have declined both proportionately and in absolute numbers. There is plenty of evidence that some former pastoralists have become village farmers, and that many village farmers have become city dwellers.[35] Actually, these shifts in residence and in economic adaptation appear to have been continuous processes throughout the history of the Middle East. There is also evidence of city dwellers moving to villages and of villagers becoming pastoral people. While some cities have, at times, been virtually depopulated, there are also areas of great stability —such as the Nile Delta and Middle Mesopotamia—which have been continuously occupied by villagers throughout recorded history.

32 Morton Rubin, "The Walls of Acre: A Study in Urban Anthropology," unpublished manuscript, (1965).

33 Charles Issawi, ed., *An Arab Philosophy of History: Selections from the Prolegomena of Ibn Khaldūn of Tunis,* (London, 1950), pp. 117–18.

34 Abdul-Razzak Hilali, *Migration of Rural Folk to Towns in Iraq,* (Baghdad, 1958), p. 25.

35 al-Wardī, pp. 258–59.

Therefore, it does not necesarily follow that all Middle Eastern city dwellers and villagers derive ultimately from pastoral ancestors.

Al-Wardī himself, in fact, questions the validity of the claims of many Iraqis that they have noble pastoral ancestors. He questions these claims on the grounds that present-day Iraqis are descended primarily from pre-Islamic, sedentary Mesopotamians, mixed with a smaller number of Arab nomads.[36] How then can he also take the position that whatever physical courage, loyalty among kinsmen, generosity, honesty and hospitality Iraqi city dwellers, and villagers too, may exhibit is is attributable to pastoral ancestors?[37] He further implies that the differences in practice among these virtues—as they occur in the desert, the village and the city—are due to the deteriorating influences of the city,[38] despite evidence that theft and murder are committed by and among nomads, and the violent factionalisms of villagers and city dwellers are seen as being essentially similar to those of the nomads.[39] These actions are not seen as being virtuous in themselves, but when linked with masculine aggressiveness they somehow acquire a heroic quality.[40]

That Middle Easterners entertain such stereotypes of themselves is in itself an important fact. Their expressions of distress in this matter may be a conventionalized emotional response to a perennial source of tension in their culture. This tension results from the conflict between repressive, corruptible governments (located in the cities) and a populace struggling to survive by means of commerce or farming or pastoralism—modes which are constantly threatened by natural and man-made disasters. The continual insecurity, and the adjustments which survival in it entails, may have found emotional release in the myth of the virtuousness of the simple desert life and its having been corrupted by uncontrollable forces in village and city life.

It is not fashionable, these days, to be caught with one's dichotomies showing. Accordingly, many scholars disavow the ones just discussed. Yet there is less inclination to disavow generalizations that "the city" is distinctive because it is sophisticated, literate, economically and politically dominant, changing, and non-traditional. Coupled with this is the tendency to explain away those city-dwellers

36 al-Wardī, p. 152.
37 al-Wardī, pp. 193, 260, 265–66.
38 al-Wardī, p. 293.
39 al-Wardī, pp. 263 ff.
40 al-Wardī, pp. 277, 297–99, 324, 336–37.

who clearly do not exhibit these characteristics by claiming that they are not "really urban" even if born and raised in the city.

It seems apparent that those who take this position are, whether they realize it or not, asserting that the only "truly urban" people are the urban elite social classes. These social classes include what remains of the preindustrial aristocracies (hardly non-traditional) plus the beneficiaries and manipulators of modern political and economic systems, Westernized science and education. It is true that such people congregate in the cities and that their presence there is one of the factors which makes the cities different from villages. But to characterize "the city" as a whole, to define it in terms of only one of its cultural components, seriously distorts reality. Cultural heterogeneity is certainly an urban characteristic, and surely an essential ingredient of urbanism is the coexistence, in a densely populated, small area, of many possible bands in the cultural spectrum. These together, not any one singly, constitute urban society.

The bias implicit in unwarranted generalizations concerning the urban elite is epitomized in the invidious comparison between refined, educated, "urban" dialects and the crude speech of country bumpkins. Doubtless such contrasts exist, but they are only one of many varieties of dialectical differences in the Middle East. Dialectical differences *within the same city* have been noted and well documented in the case of Baghdad, for example.[41] Some regional dialectical differences, too, are inter-urban rather than rural-urban, as in the case of the Tripoli dialect which is a favorite butt of ridicule in Beirut.

TWENTIETH CENTURY URBAN CULTURAL CHANGE

Urban cultural change is a congeries of processes, not a single process. It can best be analyzed in three contexts: that of migration of rural people to cities, with emphasis on their adaptations; growth of cities and patterns of change in them; and expansion of various influences from cities to rural areas, including modern innovations in technology and social systems.

Migrant Adaptations.

A sizeable portion of the people who live in Middle Eastern cities were born in villages. Indirect but massive evidence of this exists in

[41] Haim Blanc, *Communal Dialects in Baghdad,* (Cambridge, 1964).

the fact that the population of most Middle Eastern cities is increasing at a higher rate than that of the countries in which they are located.[42] Amman, for example, in 1957 contained 9.7 percent of the population of Jordan, whereas in 1915 it had only 2.1 percent of the population of the same territory.[43] The sheer magnitude of urban growth can be indicated by a couple of examples. Baghdad, whose population was about 1,100,000 in 1965, had increased almost 800 percent since 1904 when it had approximately 140,000 people.[44] The much smaller city of Tripoli, Lebanon had about 30,000 people in 1912, and had increased about 600 percent to 180,000 in 1962.[45] Similar developments have been taking place in other Middle Eastern cities. High rates of natural increase have contributed to these growth patterns, but the primary factor responsible for the excess of urban growth over nationwide growth has been the migration of rural people to the cities. For example, in 1960 more than a quarter of the population of Baghdad consisted of people who had migrated from rural areas.[46]

It might be argued that since political refugees from Palestine and economic refugees from southeastern Iraq contributed much to Amman's and Baghdad's growth, these migratory patterns are special or abnormal cases. One answer to this is that in Middle Eastern affairs, not only recently but throughout history, there has been a succession of crises which force large numbers of people to move. Another answer can be stated in terms of figures from Beirut, Lebanon. This city has been growing very rapidly, but it has not recently faced any demographic problem as acute as those which have affected Amman and Baghdad. In 1953, 26 percent of the heads of a sample of 1,933 Beirut households had been born in Lebanese villages, and, furthermore, 30 percent of the fathers of the household heads had been born in Lebanese villages.[47] While most of the migrant heads of household apparently considered Beirut to be their homes, and while nearly 80 percent of their children had been born in the city, it is clear that there has been a steady rural-to-urban migration over more

42 Gabriel Baer, *Population and Society in the Arab East,* (New York, 1964), pp. 177 ff.

43 Jane Hacker, *Modern Amman,* p. 57.

44 Abdul-Aziz Dūrī, "Baghdad," *The Encyclopedia of Islam,* (new ed.), vol. I, 907.

45 John Gulick, "Old Values and New Institution in a Lebanese Arab City," *Human Organization,* XXIV (1965), 50.

46 Al-Madfai, "Baghdad," *The New Metropolis,* p. 59.

47 Churchill, *The City of Beirut,* p. 45.

than one generation. Cairo in 1947 had a population three times the size of what it had been forty years previously, and more than one-third of these people had not been born there.[48]

Some information is available on the reasons for migrating to the city. About 65 percent of the rural migrants to Beirut had been motivated primarily by economic reasons.[49] In Alexandria, Egypt, about 60 percent of migrants from villages had been similarly motivated.[50] There is no question but that the great majority of recent migrants into Baghdad have been driven by desperate economic conditions in the villages of southeastern Iraq, and lured by the hope of greater monetary rewards in the city. Two studies which have been done recently in Turkey repeat the same theme. Twenty-five families who migrated from an Anatolian village to Istanbul between 1952 and 1960 were reported to have done so solely for economic considerations.[51] A sample of people from the squatter settlements in and around Ankara (whose inhabitants constitute one-third of the city's population) apparently migrated because of population pressure in the villages.[52] In North Africa, migrants to the city have been impelled largely by the desire to escape desperate rural conditions.[53]

How successful is the economic adaptation of rural migrants to the city? Unfortunately, there is no good information on villagers who once moved to the city but subsequently returned to the village to live. No doubt there are such people, and their impressions of city life could easily be different from those of former villagers who are now living in the city. Be this as it may, only 3.45 percent of the heads of household in the Beirut sample said that they preferred rural work to urban work, which means that most of the migrant sub-sample must have preferred urban work too.[54] Three-quarters of the sample liked their present jobs in the city—41 percent of them because of the job's economic advantages and 23 percent because of its being good

[48] Raphail Wahba, "Cairo," *The New Metropolis in the Arab World,* (1963), p. 33.

[49] Churchill, p. 49.

[50] El-Saaty and Hirabayashi, *Industrialization in Alexandria,* (Cairo, 1959), p. 72.

[51] P. Suzuki, "Encounters with Istanbul: Urban Peasants and Rural Peasants," *International Journal of Comparative Sociology,* V (1964), 210.

[52] Granville Sewell, *Squatter Settlements in Turkey: Analysis of a Social, Political and Economic Problem,* (Cambridge, 1964), p. 53.

[53] Roger Le Tourneau, "Implications of Rapid Urbanization," Leon Carl Brown (ed.), *State and Society in Independent North Africa,* (Washington, 1966), pp. 129–35.

[54] Churchill, p. 52.

training.[55] In Alexandria, many of the migrants have been able to find regular work owing to the city's industrialization, and almost 40 percent of them have been living in the city for at least sixteen years.[56] Thus, it appears that their economic motivations for moving to the city have been largely satisfied. In 1957, 57 percent of the industrial workers in Baghdad were rural migrants living in squatter settlements; they had become a major factor in the labor force, and some of them were actually prosperous.[57] They are described, furthermore, as generally preferring city life to life in the villages which they had left. Wretched as were the living conditions in the squatter settlements, they were evidently felt to be less wretched than village conditions.[58] There are similar indications from the squatter settlements of Ankara[59] and from some, but not all, of those in North African cities.[60]

The first settlers in the *gecekondu* (squatter settlements) of Ankara consisted chiefly of young married couples.[61] A little more than a quarter of the migrants living in Alexandria moved there as youngsters with their parents, but the great majority came alone as adolescents or young adults.[62] A large proportion of the recent migrants to Baghdad apparently came as adults. The point which I wish to make is that sizeable proportions of the urban populations are not only rural in birth, but in culture as well. However, migrants' personalities may undergo some change (though not necessarily drastic) after the move to the city. A projective psychological study of two groups of Algerians—one who had lived in a desert fringe oasis all their lives and the other who had moved from the oasis to the city of Algiers—contains some insights into this matter. Miner and DeVos, the authors of this study, attempted to demonstrate the changes in the personalities of migrants to the city by comparing their personalities to those of rural people.[63] As I have pointed out elsewhere, "the personality differences between them, as revealed by the 'blind' analysis of the Rorschachs by DeVos, are not striking except in the urban peoples' greater intrapsychic tension and hostility which are attribu-

55 Churchill, p. 52.
56 El-Saaty and Hirabayashi, p. 74.
57 Hilali, p. 89.
58 Hilali, p. 63.
59 Sewell, p. 110.
60 Le Tourneau, pp. 129–30.
61 Sewell, p. 77.
62 El-Saaty and Hirabayashi, pp. 72–73.
63 Horace Miner and George DeVos, *Oasis and Casbah: Algerian Culture and Personality in Change*, (Ann Arbor, Michigan, 1960).

ted to the aggravated pressures of the French in the city."[64] As might be expected, those migrants whose personalities showed rigidity to start with had more difficulty in adjusting to city life than did those with less rigid personalities. Over and beyond the rural-urban differences which were discovered, these authors conclude that "the personality type which marks the Arabs as a group rests on the common developmental experience which the culture dictates."[65]

Large portions of the inhabitants of Middle Eastern cities were once migrants. Economically they are substantially integrated into the life of the city. If this integration has been largely achieved at a severe emotional cost, there is very little documented evidence of it. Le Tourneau does declare that the North African farmer or shepherd who has migrated to the city finds the adjustment to the "continuous and tyrannical rhythm of the machine" very difficult, and he even attributes the scarcity of skilled factory workers to this difficulty.[66] The fact remains, however, that relatively few migrants in North Africa or elsewhere in the Middle East are faced with having to adjust to industrial work rhythms. The rural migrants bring their culture with them to the city, and there are a number of ways in which their cultural patterns share aspects of city culture. One of these ways is the contribution of rural migrants to the socio-culturally distinct quarters which are a feature of Middle Eastern cities. For obvious reasons, many new migrants must depend on the assistance of kinsmen and friends who are already established in the city. When the latter find the newcomers a place to live, it is likely to be close to their own residences. This results in many residential clusters of people who originally came from the same village or the same rural region.

In Baghdad at the present time, three of the quarters in Rusafa, the old city on the east bank of the Tigris River, have distinctly tribal names. As far as I could tell, they are no longer primarily inhabited by members of the tribes in question. They are located in what is now the midst of "downtown" Baghdad, which is gradually changing from an old-style Arab city of narrow, twisting alleys and covered *sūqs* to a modern business and industrial district. These particular quarters, and some others, are still primarily residential, but they are surrounded by cultural transformation. One hundred years

64 John Gulick, "Review of Miner and DeVos, 'Oasis and Casbah,'" *American Anthropologist*, LXIII (1961), 858.

65 Miner and DeVos, p. 186.

66 Le Tourneau, p. 139.

ago these same quarters were on the outskirts of what was then the whole city of Baghdad, and they were inhabited by newly arrived tribal groups from rural areas. The names may well be changed in time, but so far they have contributed to the character of the city. This same pattern was evident in the internal structure of the *sarā'if* (squatter settlements), which by 1960 had grown to massive proportions on the outskirts of Baghdad. Mostly farmers from the province of Amara in southeastern Iraq, the population of the *sarā'if* comprised eleven tribal groups. They were established so that the members of each group were clustered together. Their shaykhs continued to exercise their authority by adjudicating disputes and maintaining guesthouses. While tribal codes of conduct and indemnification were maintained, they were adapted somewhat to city conditions and problems such as the one created by the operation of motor vehicles.[67] When most of the *shurugiyya* (squatters) of Baghdad were forcibly relocated by the government in 1963, these tribal clusters were broken up, and I do not know to what extent they were re-established.

Similar behavior has been reported from Turkey. Suzuki describes how the members of an Anatolian village living in the *gecekondu* of Istanbul kept very much to themselves, maintained village endogamy, and established their own mutual fund covering funeral expenses, transportation fares to and from the home village, and help for newcomers.[68] In the *gecekondu* of Ankara, there are definite village-of-origin clusters, although there are also many households which are not so affiliated.[69]

These clusterings are by no means limited to the mushrooming squatter settlements of modern Middle Eastern cities. For example, in the Bāb al-Tibbānī section of Tripoli, Lebanon, there is a residential cluster of Alawites from Jabal Ansāriyya in Syria, and in the Qubba section of Tripoli there is a community of Maronites all from the same village high on Mount Lebanon. Physically these two clusters are integral parts of the city. Their inhabitants maintain village ties (and, significantly, each cluster is close to the most convenient transportation facilities by which the villages can be reached), and to some extent they operate their own shops, but in other ways they are parts of the city itself. Each cluster has a definite reputation for certain kinds of behavior, and for preferring certain kinds of work. Ob-

67 Hilali, pp. 62–71.
68 Suzuki, "Encounters with Istanbul," *International Journal of Comparative Sociology*, V (1964), 210–12.
69 Sewell, p. 180.

viously they contribute to the very essence of Tripoli's "urban heterogeneity." In Cairo, the locations of over one hundred village mutual aid societies have been analyzed. They tend to be close to the bus stations used by travellers to and from the home villages. This suggests some residential clustering by village.[70] Village mutual aid societies in Alexandria have also been identified.[71]

Maintenance of village ties naturally implies communication,[72] but beyond this it means a number of other things: 1) Regular returns to the home village for weddings, funerals, and other ceremonies;[73] 2) regular exchanges of gifts (such as home-produced food from the village) and of visits; 3) seasonal shifts in residence (a third of the Beirut sample households went to villages in the summer; almost 5 percent of the Beirut-dwelling heads of household actually did work in villages during part of the year[74]); 4) commuting on a monthly, weekly or even daily basis.

I have observed frequent commuting in two instances in Lebanon. In one case, people from the village of Munsif travel easily by car to Beirut in only slightly more than an hour. In the other, people from villages in the Kura district can travel to Tripoli in less than half an hour. To the extent that such people have sources of help and support in readily accessible villages, they may not need to establish mutual aid societies in the cities. Indeed, I have not heard of any such societies in Lebanon where distances are relatively short. There may be more necessity for them in Egypt, where travel between village and city can be very lengthy and difficult. Also, in Tripoli I observed that the Kura villagers did not live in any residential clusters in the city, whereas the Alawite and Maronite villagers—both farther from their villages—did. Another difference, however, was that of social class. The Kura villagers were primarily white collar and professional people, whereas the Alawites and Maronites were primarily laborers and small shopkeepers.[75]

Villages which are very close to cities require some further comment. They are not (in the Western sense) suburbs which have been

[70] Janet Abu-Lughod, "Migrant Adjustment to City Life: the Egyptian Case," *American Journal of Sociology*, LXVII (1961), 22–32.

[71] M. Sedky, "Groups in Alexandria, Egypt," *Social Research*, XXII (1955), 441–50.

[72] Laura Nader, "Communication between Village and City in the Modern Middle East," *Human Organization* XXIV (1965), 18–24.

[73] Gulick, *Structure and Change in a Lebanese Village*, pp. 91–92, 98.

[74] Churchill, *The City of Beirut*, pp. 49, 53.

[75] Gulick, *Tripoli*, p. 182.

created by city dwellers wanting to live outside of the city. Rather, they are peasant villages, or part-peasant villages, which are so close to the city that they can be called satellites of it. In the Karrada section of Baghdad, which forty years ago consisted mostly of date-palm plantations, there is a village which has now physically been engulfed by the city. There remains a somewhat localized core of clansmen where the village used to be. On the outskirts of Baghdad, there is a still separate village named Saydiyya, which, twenty years ago, was still primarily a peasant village. By 1964, however, less than a third of the households were still involved in farming. The majority had found jobs in the nearby Dawra refinery and other plants. While the non-farmers seemed more receptive to innovations than the farmers, the two groups still had much in common as far as their cultural patterns were concerned, and both were being equally bombarded by the mass media and other expressions of the changing world.[76] Another case is Būrrī al-Lamāb, a village of 2,400 people located five miles from the central market of Khartoum, Sudan. Most of its inhabitants are no longer farmers, but work in the city. Its inhabitants also include city workers who have moved out to it, indicating a truly suburban development.[77] For many residents, apparently, the village is "hardly more than a bedroom," and they are accustomed to using city dress and manners in Khartoum and village ones in the village.[78]

A third case is the village of Danizköy, on the Sea of Marmara near Istanbul. The villagers are thoroughly adjusted to a cash economy. Those who live in Istanbul are not particularly village-centered in their attitudes, and yet many of them have returned to the village after a sojourn in the city.[79]

These are villagers who do not have especially localized sentiments, apparently owing to their very close involvement with the city and the national culture.

A fourth case is the highly complex network of standing interrelationships, and the "hundreds of minor daily interactions" by which those interrelationships are carried out, which link the hundred-odd villages and towns of the Kerman Basin in Iran with the city of Ker-

[76] Najia Hamdi, "Some Aspects of Social Change in Saydiyya, a Semi-Industrialized Village near Baghdad," (American University in Cairo, 1964), Unpublished M.A. thesis in anthropology, p. 58.

[77] Harold Barclay, *Buurri al-Lamaab, A Suburban Village in the Sudan,* (Ithaca, 1964), pp. 12, 15.

[78] Barclay, p. 269.

[79] Suzuki, "Encounters with Istanbul," pp. 212–13.

man.[80] The existence of such villages further supports the idea that in the Middle East, village and city subcultures have much in common.

URBAN GROWTH

In common with other parts of the non-Western world, the Middle East has a small number of primate cities which seem, on various standards, to be disproportionately large. Cairo, the biggest by far, now has about 4,000,000 people; Alexandria and Baghdad have about 1,000,000 apiece. Each of these cities' populations has increased about 500 percent in the past fifty years. This is also true of the several somewhat smaller primate cities, and, on the whole, of the many more secondary ones.

Such growth is in itself an important change. In various ways, however, the impact of that change seems to be somewhat softened. Rural immigrants tend to create their own institutional buffers based on village models. And from the traditional urban side, Professor Abu-Lughod points out in her paper that about 70 percent of the people in Cairo live in a style of life which is neither "modern urban" nor supported by rural-model buffers. If this is true in Cairo, it is probably equally true in smaller cities. In some of the latter, certainly, the kinship-based, individualized-personal patterns persist strongly.[81]

Western technology has greatly changed the transportation, communication and public health aspects of Middle Eastern cities. Coupled with international political and economic interests, it has also made possible many new types of business and public institutions —of which the public school is one of the most important. Daniel Lerner has characterized "moderns" as those people concentrated in cities who seem to have most adapted themselves to these innovations and whose explicit expectations would appear to be in conflict with most of the general cultural patterns which I have outlined.[82] But such people are still outnumbered in the cities, as well as the country, by "traditionals" and "transitionals," to use Lerner's own terms. These latter apparently maintain the older patterns as best they can. First, most of the new urban housing tends to perpetuate the traditional combination of residential and commercial functions. Second, the suburban commuter settlement of the West is present only

80 English, *City and Village in Iran,* pp. 67–68.
81 Gulick, "Lebanese Arab City," *Human Organization,* pp. 49–52.
82 Lerner, *The Passing of Traditional Society.*

on a very small scale in the largest cities and is virtually absent from the others. Third, culturally distinct quarters, though changing, are neither obliterated by industrial growth nor "bureaucratized" out of existence by conversion into mere administrative units. For example, Basta, the Sunni Muslim quarter of Beirut, does not coincide with any census tract or ward boundaries. To the extent that the immigration of rural people causes cities to grow, it may also contribute to the continuation of "traditional" urban cultural patterns. However, there now are some indications that the rate of natural population increase in cities is becoming higher than the rural one. If this new trend, which has been observed in Egypt,[83] becomes general, urban natural increase may in the future become a more important factor in urban growth than immigration. This, in turn, may accelerate modernization in urban culture.

RADIATION OF INFLUENCES FROM THE CITY

The radio is an unprecedented medium of urban-to-rural communication, and television is now being added to it in many places. The role of the burgeoning press is not clear, owing to the consistently high rates of illiteracy especially in the rural areas. However, the establishment of uniform national school systems is a major goal of nearly all Middle Eastern countries. As this goal approaches realization, the number of illiterate migrants to the city will become smaller and smaller. Concurrently, if industrial and administrative institutions and functions become more decentralized than they are now, the impetus for migration to the cities itself may be lessened.

There are now signs in the planning policies of Middle Eastern countries that these possible trends may become realities. I leave it to others to wrestle with the semantic problem of whether these trends will make the Middle East more urbanized than it is presently.

DISCUSSION

The anthropologists began the discussions by raising some question as to the appropriateness of the overall methodological approach. Professor Fernea was disturbed by what he regarded as Professor Gulick's static approach: "When we speak about urbanization we're

[83] J. Abu-Lughod, "Urban-Rural Differences as a Function of the Demographic Transition: Egyptian Data and an Analytical Model," *American Journal of Sociology*, LXIX (1964), 476–90.

talking about a *process*, and cataloging differences and similarities does not necessarily tell us how to get into questions of social change and social process."

Professor Nader stressed the importance of a comparative approach to such a problem: "Having worked in both Mexico and Lebanon, I was very struck by the degree to which there are similarities between urban and rural people in Lebanon while they are absent in Mexico. In Lebanon there are lines of communication between urban and rural people which are not present in Mexico. Such communication is found in Greece, but not in the Netherlands. To understand such relationships we have to consider why there is communication in some areas but not in others. The first part of Professor Gulick's paper struck me as a description of Arab national character. To understand similarities between rural and urban peoples we need some tighter controls in order to find out exactly how the city affects people who are either recent migrants or natives, as opposed to rural people."

Professor Issawi, viewing the problem of similarities and differences between the city and the countryside in historical perspective, observed that "it is not just a question of urbanization, but also of Westernization. As the cities become Westernized, the countryside does not. This probably increases the gap between city and country in some respects, certainly the gap between the elite of city folks and the rural people without any education."

Professor Issawi's remarks led Professor Safran to point out that in comparing city and village, a considerable difference, not similarity, must exist in literacy rates and exposure to mass media. Professor Abu-Lughod added, "in my opinion, the single most important characteristic defining different populations in Egypt is the literacy rate. In a comparative analysis of Egyptian towns, I set up a scale for scoring urban communities. I discovered that, especially for higher education, I could throw away the scale and retain the literacy rate alone —particularly the female literacy rate—as an index of the kind of place involved."

Professor Gulick remarked that though this is true, he thought that the tendency of modern education is to reduce this difference. He cited the construction of village schools. However, effective literacy, rather than the enrollment of students in schools, is most important. He added, too, that as far as communication is concerned, the radio, and even TV, have become important factors.

Professor Grabar saw some difficulty over the meaning of literacy: "Whatever the technical literacy rates, (judging from my work of several seasons in the central Syrian desert) I was struck by the gap between cities and the hinterland. There, schools are built, but the gap between Damascus and Palmyra and Damascus and Sukne or Taybay has, if anything, widened. They hear the radio, but no newspaper reaches that area. Palmyra may get a fifth-rate Homs sheet, but will not get a Damascus newspaper unless somebody brings it on the airplane. No newspaper gets to Sukne, Taybay and other points north and northwest. Books, which in any city are sold on the sidewalk, are unheard of in these places. Even though literacy in the sense of being able to read and write does increase, people don't read and write, or don't read the same things."

Professor Adams questioned the significance of literacy as an index of city-country differences or of social change: "I think a certain academic bias crept into our discussion of literacy. What counts about literacy is not the ability to communicate on a world scale, or to communicate with various kinds of elites; rather, the crucial issue is one's ability to participate in the political process and to participate in urbanization as a social and cultural phenomenon. In these terms, the functional requirements for literacy have very sharply declined with the gradual erasing of dialect lines through the spread of radio, and even of television, which I think is going to be increasingly a factor throughout the Middle East. If this is so, the rate of professed literacy may continue to be useful as an indication of much broader cultural patterns. But I don't think we can assume that literacy—that functional literacy—is going to be a requirement for increasing participation in the political process."

Professor Fernea took up in detail some of the "trait-complexes" discussed by Professor Gulick, and commented further on some of the difficulties involved in using indices of this kind. Considering household size, he observed that a student of his did a study of Port Said slums which compared the housing projects which the government constructed after 1956 with the slums that still existed in the city: "This student (al-Sibai; see footnote 19, p. 135) found an amazingly diverse collection of people in the same apartment—people renting out sleeping space by the square foot. Any census of housing units might have found the same large numbers as live in large rural families, though the people living in the Suez apartments were neither kinsmen nor even people from the same rural areas of the country. We know too from studies of Nubians in Cairo that it is the custom of Nubian men to rent an extra room in a house and live with other families. I suspect that the composition of a household in an urban setting, especially in the lower income groups, is far more diversified than the composition of a rural household.

"Another point which I think deserves some further attention is the factionalism found in both the urban and rural segments of the society which Professor Gulick sees as a factor in the similarity of the two areas. It seems to me that the basis for rural and urban factionalism is usually, if not always, quite different. In my experience, the basis of factional cleavage in villages is almost always kinship. A few villages in Egypt are divided between Muslims and Copts. Thus there are cleavages along religious lines too, although client-patron relationships between Copts, when they are in the minority, and Muslims, protect the village structure from disruptive religious factionalism. In the city, it seems that factionalism is based largely on such sectarian differences, and more recently, on political and class differences. Some of the revolutionary movements in the Middle East and the riots seem to have elements of the "haves" and "have-nots" about them."

Professor Lenczowski also drew attention to the fact that "in recent times, especially following the revolutionary movements in some of these countries, politics superimposed a new factionalism in the cities upon the old patterns. Consequently, in Baghdad you can quite clearly determine which is the Ba'athist area, which is the "Iraq-first" nationalist area, and which is the Nasserite area. Sometimes this has a connection with the old divisions of the city according to either tribal, ethnic, or religious divisions, but sometimes it does not. In one region, for example, there is a very strong Communist element in the most fundamentalist Shī'a area."

Professor Gulick, as he indicated in his paper, thinks this contrast between city and village types of factions is overdrawn. He believes that sectarian and other non-kinship factors are as much at work in villages as in cities.

Professor Fernea turned to another point on which his judgment differed from that of Professor Gulick. Professor Gulick, he remarked, "observes that peasant and city-dwellers are alike in their awareness of sectarian differences and in exclusiveness of behavior, yet it seems to me that this is one of the things that is changing in the Middle East. Sectarian differences in some Middle Eastern cities, and in some elite circles, are being replaced by perfórmance criteria, which to some extent makes it possible for people to be placed according to abilities rather than according to membership in some group of larger definition. I realize this is only in process, but I wonder if sectarian identity will not cease to be a basic factor in occupational and political institutions."

Professor Lenczowski, on the other hand, did not see a change from sectarian or communal considerations to performance criteria as essential to the modernization of the Middle East: "This undoubtedly is happening, but it is not a clear-cut one-way movement. I think that there are at least two possible courses for modernization to take: one in which the achievement criterion with its attendant social mobility prevails; and the other, as exemplified in Lebanon, where a political balance of one sect against another, or one ethnic group against another, is maintained. The Lebanese with their considerable political realism have come to the conclusion that the discrepancies in communities' achievements cannot be eliminated too easily. The various communities must have adequate recognition in the political process and distribution of spoils, and religious and ethnic divisions are expected to produce an adequate balance."

Professor Fernea then turned to the obverse question, the impact of cities on rural areas, which Professor Gulick mentioned briefly in conclusion. From another point of view, here we see the operation of forces which tended to bring urban and rural areas into closer resemblance, though not, as in the case of migrants, by the formation of "rural" populations in the cities, but by the "urbanization" of the countryside. Professor Fernea observed that "when the migrant comes to the city, he is isolated, by and large, if not physically in some separated areas like the *ṣarīfa* regions around Baghdad, at least in a lower-income area where his opportunities to observe and to come into personal contact with people of other socioeconomic classes, and people from other areas or people who are native to the city itself, are limited. On the other hand, in countries which have experienced revolutions in the last couple of decades there is a decided increase in the number of city administrators who are being sent to country villages to open a variety of social services. Unprecedented in the past, this brought urban people into meaningful contact with rural people. The rural people have a chance to observe the city people living their daily lives. They gossip about one another. They learn about each other. They come into contact with each other formally and informally, to some extent, and it seems that this is one of the areas in which meaningful role models, different from those the rural person has traditionally experienced, are transmitted. The schoolteacher who causes the tribal girl to aspire to become a schoolteacher herself, the doctor in the village who makes the isolated tribal boy think about becoming a doctor, or leads him to take advantage of some of the educational opportunities now being made available—such people are crucial. If we are looking for the areas in which change creates commitments common to both rural and urban people, and even perhaps life styles common to both groups, I suspect that villages such as these are just as important as the city—granting that the city is the ultimate source of new ideas."

JANET ABU-LUGHOD

Varieties of Urban Experience: Contrast, Coexistence and Coalescence in Cairo

THEORETICAL PERSPECTIVE

The urban-rural antithesis, which has long preoccupied American sociologists and anthropologists and has provided many of the false premises from which the effects of urbanization have been deduced, has become increasingly untenable as more varieties of urban experience come within our purview. Not only does the content-meaning of "rural" and "urban" shift from one technological era to the next and from one cultural context to another, but the universally visible contrast between hamlet and city, even of one time and place, often blinds rather than illuminates. The resulting optical illusion encourages an unwary substitution of "urbanite" for city, "villager" for village. Few simplifications can cause greater havoc in comprehending the Middle Eastern city than this all too facile identification between person and place.

The contrast between city and village is certainly as great in Egypt as anywhere. A brief trip from Cairo to the countryside transports one to another world. Yet even within Cairo itself one may pass within minutes from elegant hotel to rural-style mudbrick hut, a fact which suggests that, at a minimum, two varieties of "urban" life coexist in the capital. It would be an error, however, to deduce from this that only recent rural migrants live in the "villages" of Cairo or that "urbanity" is a natural attribute of the city-born.[1] Not only

[1] My earlier attempt to investigate this question was reported in "Migrant Adjustment to City Life: The Egyptian Case," *The American Journal of Sociology*, LXVII (1961), 22–32. The predominant sites of migrant first settlement up through the early 1950's appeared to be located within an elliptical ring surrounding the central business district, rather than in the more rural peripheral areas. It was hoped that direct data on where migrants settle within the city would be made available in the census of 1960, but unfortunately, the cross-tabulations required for this analysis were again omitted from the census processing.

would these generalizations be inaccurate but, even more important, they would be seriously incomplete. If one added together all sections of the city clearly modeled on the rural prototype and all sections resembling modern western cities, one would still be left with most of the city unclassified. Any conceptual scheme that does so inefficient a job of taxonomy needs revision.

In Cairo one may move not only across a "scale" dimension of urban–rural but across a "time-technology" dimension as well. Side by side stand the modern factory and the primitive workshop, the bank and the turbaned moneylender, suggesting the persistence of a vital residue from yet another variety of urban living. This "time and technology" dimension, sometimes described as the traditional-modern antithesis, cuts diagonally across the urban-rural continuum. Without defining these terms more precisely at this point, we can set up a very simplified matrix of four "ideal type" life styles, any and *all* of which might be found within an administrative territory defined as "urban."

TABLE 1
A Simplified Typology of Life Styles

TIME–TECHNOLOGY

Ecological Scale	*Traditional*	*Modern*
Small	TYPE 1 villager	TYPE 2 modern small-town dweller
Large	TYPE 3 traditional urbanite	TYPE 4 modern urbanite

An important *caveat* must be inserted here. This typology is best applied within a given culture at a given time; serious questions exist as to whether any typology can be found that will be universally applicable across cultures and across long stretches of historical development. As the character of society changes, the elements distinguishing urban from rural patterns will alter and the "modern" of yesterday may become the "traditional" of tomorrow. Thus, when we describe the contemporary manifestation of "traditionalism" in Cairo, this is not to be confused with an unchanging heritage of the past. We must treat the variables of urban-rural and modern-traditional as parameters of a moving range rather than as invariant polarities in an ideal system.

The early theories concerning urban-rural differences viewed the characteristics of villagers and urbanites essentially as consequences of the scale of the community. Attention was focused almost exclusively upon contrasting traditional villagers (Type 1) with modern urbanites (Type 4). The effects of urbanization upon cultural patterns and the adaptive requirements of rural-to-urban migration were predicated on the basis of this two-cell model. Only belatedly has the analysis of American urbanization shifted its focus to Types 2 and 4, in recognition of the fact that examples of Type 1 are increasingly difficult to locate except in anachronistic "poverty pockets," whereas Type 3 is now confined to the declining enclaves of "urban villagers" being swept away in the wake of renewal projects.[2]

It is Type 3, however, that constitutes the missing link in Cairo's puzzling ecology. Even today it accounts for more than half of the city's population. Failure to conceptualize adequately this third type—not as some intermediate point along an urban-rural continuum but as a separate dimension—accounts for the frequency with which irrelevant questions and meaningless hypotheses are framed about the Middle Eastern city. Whether we wish to account for past changes or to predict the consequences of today's changes, this third element must be taken into account.

The task of this paper is to clarify the nature of these three variations on the urban theme—the rural, the traditional urban, and the modern—as they appear in Cairo, to demonstrate their proportional distribution and physical disposition within the contemporary city, and to suggest some of the underlying factors that not only have given rise to the coexistence of these three urban worlds but which continue to shape their coalescence. In so brief a discussion, however, we can do little more than sketch the major outlines.[3]

2 The term "urban villager" has been borrowed from a study of an Italian-American ethnic enclave in Boston. (Herbert Gans, *The Urban Villagers*, [New York, 1962]). While the term is expressive, it should be considered aphoristic only, since the life style it describes approximates more that of the traditional urbanite than either the modern city dweller or the villager.

3 A detailed and documented account of the city's lengthy history and its evolving pattern can be found in J. Abu-Lughod, *Victorious City: The Growth and Structure of Modern Cairo* (Princeton University Press. Forthcoming). A complete description of the statistical methods used to study the ecological structure of contemporary Cairo and of the specific findings derived from these methods is included in J. Abu-Lughod, "The Ecology of Cairo, Egypt: A Comparative Study Using Factor Analysis," (Doctoral Dissertation, Department of Sociology, University of Massachusetts, 1966). Data on each of the 216 census tracts in Cairo in 1947 plus a key to their locations are included in J. Abu-Lughod and E. Attiya, *Cairo Fact Book*, (Cairo, 1963). Unless otherwise noted, supporting evidence is to be found in these sources.

HISTORICAL PERSPECTIVE

Before considering contemporary Cairo, let us first step back to gain some social and historic perspective. From this vantage point four items stand out above the mass of many absorbing details.

The first is the vast scale of the metropolis and its rhythm of growth. In the fourteenth century, Cairo was the most populous city of Europe and the Mediterranean basin, containing along its extended north-south axis almost 500,000 inhabitants. By the eighteenth century her population was estimated at only 300,000 and this figure had declined to 260,000 by the turn of the nineteenth century. A few decades later it reached a temporary nadir of about 200,000. It has, therefore, been during little more than the past century that Cairo experienced a new cycle of growth, expanding in population from some 400,000 in 1880 to 600,000 by the turn of the twentieth century, topping 1,000,000 by 1927, 2,000,000 by 1947 and exceeding 4,000,000 today. In 1960, with a census total of slightly under 3,500,000, Cairo was the twenty-third largest city in the world and the *only* city of the Mediterranean, Middle East, or Africa to have passed the 3,000,000 mark.[4]

This recent upsurge is the result of two trends: first, a demographic revolution within Egypt; and second, a rapidly increasing rate of urbanization. From the middle of the nineteenth century, when Egypt's population began to mount steeply from its "underpopulation" starting point, Cairo's rapid rate of expansion paralleled that of the country as a whole, a situation which prevailed until only a generation ago. Since Egypt's passage over the threshold of industrial modernism, however, Cairo's population has been growing about twice as fast as her supporting rural base, so that the city now houses one out of every 7.5 Egyptians. Each year it experiences a net increment of at least 100,000 persons—sufficient to populate a good-sized city. In terms of the future, we may reasonably expect a city of 7,000,000 inhabitants within a decade, if not sooner.[5]

[4] For a complete listing of the twenty-four cities with more than 3,000,000 inhabitants, (five in North America; seven in northern Europe; four in Central or South America; seven in the Far East), see Peter Hall, *The World Cities*, (New York, 1966), p. 11.

[5] This projection is based upon a simple semi-logarithmic projection of the most recent growth trends in which migration has been compounded by a sharp upturn in the Cairo rate of natural increase. Unless urban birth rates decline rapidly, this new phenomenon will inflate Cairo's population for some time to

Cairo, then, is no mean entity to be comprehended impressionistically from random observations. The findings we shall present below are based upon careful analysis of a wide variety of demographic, social and economic measurements computed from the censuses of 1947 and 1960 for each of the more than 200 administrative subdistricts (*shiyākhāt*) into which the city is divided.[6] While some of these districts are rural and others have the characteristics of modern urban communities (albeit somewhat transfigured by culture), most are a complex compound of traditional and modern aspects of urbanism, differently proportioned.

The long history of Cairo and the recentness of her rebirth, however, present a paradox: here is a city with a significant past stretching back more than a thousand years, but here also is a physical entity most of which, although appearing prematurely aged, was actually built yesterday. This is the second point we wish to stress. The Cairo mapped by Napoleon's savants soon after 1798—which was little changed from the Cairo of the fifteenth century—covered only some five square miles at its core. It consisted of tenth century-founded al-Qāhira, plus two miniscule port suburbs at Miṣr al-Qadīma (heir to the defunct seventh century town of Fusṭāṭ) and Būlāq (a post-fourteenth century addition on a newly linked sand bar). The Cairo of today, however, covers more than seventy-five square miles. Diversely urbanized, it is a vast sea in which the direct medieval heritage surfaces spottily in only three small "islands."

Herein lies the paradox. The ecological development of the immense contemporary city is anchored to these dissolving "islands" and cannot be comprehended without reference to them. What is more, each of the urban worlds within Cairo is permeated somewhat by the ether of the past that exudes from them. The tradition of premodernism persists in large parts of the city. What forces keep it alive, since the direct physical heritage is so circumscribed? It is our contention that the economic organization of the city continually reinjects, re-

---

come. The possibilities of an eventual decline in urban birth rates are explored in J. Abu-Lughod, "The Emergence of Differential Fertility in Urban Egypt," *The Milbank Memorial Fund Quarterly*, XLIII (1965), pp. 235–53.

6 The boundaries of Cairo altered drastically between 1947 and 1960. In our analysis we have retained the boundaries of the earlier date and replicated as closely as possible the same census tracts, whether or not they were officially within the city limits in 1960. Thus our study includes the west bank of the Nile, despite the fact that this area was redistricted to the Province of Giza, and eliminates the satellite city of Ḥalwān, even though this is now a part of the Governorate of Cairo.

vitalizes and reinforces such traditionalism. This brings us to the third point.

Before 1800 and even throughout most of the nineteenth century, despite the sporadic efforts of Muḥammad 'Alī to introduce industrial production and of Ismā'īl to mechanize agricultural processing, and despite some conversion to specialized cash crops, Egypt had a single economy in which the methods of production, the type of division of labor, and the social institutions relating "labor" to "management" could all be described, in retrospect at least, as traditional.[7] Even a sharp break between the urban and rural economies was absent, since urban wealth was still largely dependent on rural returns and the cities themselves were populated largely by agriculturalists. As a measure of the latter we may use Ismā'īl's admittedly inaccurate Muqābala survey of 1877, in which some 57 percent of Cairo's economically active population was listed as engaged in farming;[8] the remainder were absorbed in traditional crafts, administration, personal services or brute labor, since the modern sector had not yet appeared.

By the end of the nineteenth century, a tiny incipiently modern sector had been precariously superimposed upon the dominant traditional one. However, its impetus (primarily European speculative capital) and personnel (European and non-Muslim Egyptian) assured that it would remain insulated from the traditional economy. Far more important was the conversion of Cairo's economy to a more "urbanized" base. By 1907 these changes were already apparent in the labor force statistics of the Census. Less than 10 percent of male Cairenes were still farming full time[9] and a new modern, although still only "semi-industrial," sector had emerged which engaged the energies of some 15 percent of the male labor force.[10] Within this

[7] Albeit with market and exchange sectors.

[8] Figures giving the occupations of Egyptians (presumably females as well as males, since the "active" population so classified was almost half of the total population) by administrative unit of residence as of December 31, 1877, appear in Egyptian Ministry of Finance, *Essai de la statistique générale de l'Égypte,* (Cairo, 1878), Tables LIV and LV, pp. 115–21. Of all the urban governorates, Cairo had the highest proportion of its labor force in agriculture, probably a function of over-generous boundaries. It should be pointed out that Ismā'īl's "census" was an undercount and therefore must be used with caution.

[9] Eliminating females from the labor force data in Egypt of itself creates some decline in the percentage of agriculturalists, but this would not be sufficient to account for the drastic decrease in full-time farmers within the generation covered here.

[10] *The Census of Egypt Taken in 1917,* (Cairo, 1921) contains a "Comparative Table of Occupations by Sex, Governorates and Provinces for 1907 and 1917,"

sector, however, a disproportionate majority were foreign born; most native Egyptians were relegated to servile capacities.

Ten years later—at the height of colonial power when foreigners constituted 10 percent of the city's population and monopolized the modern professional, commercial and technical occupations—the agricultural sector had declined even more, accounting for only 5 percent of the male labor force of the city. The modern sector, still small and foreign-dominated, absorbed no more than a quarter of the male labor force.[11] In brief, according to our estimates, most of the occupied males of Egyptian origin in Cairo in 1917 were more or less integrated into a pre-industrial but nevertheless urban economy, in which the scale of firm was tiny, the division of labor organized by product rather than process, capitalization, mechanization and inventories all minimal, and the relationship between client and producer, employer and employee direct, personalized even when fleeting, and encrusted by the generosities and cruelties of the non-cash nexus.

Since the 1920's, when Egyptianization of the modern sector began, one notes a steady expansion of the modern, a stabilization and then decline of the traditional urban sector, and a continued but undramatic attrition of agriculture. Our estimates from the most recent census suggest that by 1960, 40 to 45 percent of the male workers in Cairo were in occupations or industries that might generously be defined as "modern," either in the sense of being new functions or of utilizing modern processes of production.[12] The remainder still

(Volume II, 380–404) relevant for Cairo. We have used this table to estimate the male labor force base of the city in the two census years. The estimate of modern versus traditional was made by classifying each occupational-industrial category used by the census (there were 196 entries for various activities) as either predominantly traditional or predominantly modern and simply summing the males engaged in each. While admittedly this procedure contains errors arising from both data inaccuracies as well as subjective judgment, it represents at least a preliminary attempt to estimate a magnitude which can never be measured precisely.

[11] *The Census of Egypt taken in 1917*. Additional information supporting the contention that the modern sector was largely foreign can be found in other tables in this same volume which cross-tabulate nationality with occupation, with industry classification, and with functional role (employer, employee, etc.). Processed tables derived from these cross-tabulations are available from the author.

[12] We have tried to follow the same procedure in estimating the 1960 breakdown but, since the occupational categories in the *1960 Census of Population* (Volume for Muḥāfaẓāt al-Qāhira, Cairo, 1963) differ somewhat from those included in earlier censuses, and since the technologies within specified categories have in some cases altered radically, a new set of dichotomized judgments was required. Comparability is thus somewhat reduced.

worked at traditional occupations or in industries, enterprises, and services run along traditional lines. By then the number of full-time agriculturalists was minimal and the role of foreigners insignificant.

Viewed from one side, this is an impressive transformation indicative of the rapid strides in "indigenization" and modernization that have already been made. Viewed from the standpoint of the nature of the city, however, it is striking that more than half of the occupied males in Cairo are still within the traditional sector of the urban economy, even though the organization of that sector is gradually increasing in scale and capitalization.[13] The persistence of the dual economy both reflects and intensifies a deep cleavage in the urban worlds of Cairo, in many ways far more significant today than fifty years ago when the modern sector was merely "foreign" and hence extraneous. The rent now cuts sharply into the heart of Egyptian urban society but this is, in itself, a hopeful sign pointing to eventual integration; at least now the conditions of entry into the modern sector are potentially attainable by all.

In the foregoing remarks, occupation was used to denote something broader than economic role. It was used as an index to a "way of life." While there is no absolute correspondence between an occupational role and the characteristics of those who fill it, nevertheless occupations do "select" differentially and then serve to condition men, not only by their work experiences and contacts but by their remunerations which permit further differentiations in life styles. This is true not only at the professional level but in the lower echelons of industry and commerce as well. The conditions of entry into the modern sector of the Cairo economy vary but include a minimum of literacy for jobs above unskilled labor.[14] This in itself is a distinguishing element in *Weltanschauung*. Furthermore, the fact

[13] The average number of workers per firm is one way of measuring the "scale" of enterprises. By this measure, Cairo remains dominated by a preindustrial level of scale, for as recently as 1957 this *average* size was still approximately three, despite the existence by then of many large-scale firms. Our computations are from chapter 19, p. 400 of the U. A. R. Government's official *Annuaire Statistique 1960 et 1961*, (Cairo, 1962).

[14] For example, in a study conducted in 1954 among a sample of workers in large factories in Alexandria, it was found that 75 percent of the urban-born employees and 63 percent of those born outside cities (who held, on the average, the less skilled jobs) were literate, even though the literacy rate for Alexandria as a whole, including the middle and upper classes not found in factories, was only 50 percent for males. See the findings reported by Hassan el Saaty and Gordon Hirabayashi, *Industrialization in Alexandria*, (Cairo, 1959), pp. 123–24. More than 90 percent of those workers earning L. E. 4 per week or more claimed literacy.

that industrial establishments in the modern sector are, on the average, quite large means a greater exposure to secondary relationships for their workers than occurs in the family-size firms that dominate the traditional sector. It also requires greater adjustments on the worker's part to sets of bureaucratized regulations and procedures. Even more important, this larger size affects average wages. Other things being equal, the larger the scale of an enterprise in Egypt, the more modernized the industry and the higher the average wages.[15] While all these factors do not *ipso facto* "create" modern attitudes and behavior in personal life, they cannot but help to further differentiate Cairenes along the dimension of traditionalism-modernism, even in non-economic areas of life.

The final point we wish to raise concerns migration. If the urban economic structure helps to sustain traditionalism within Cairo, it is the equally potent force of migration that helps to infuse it afresh with ruralism. Migration from the countryside, however, is as ancient as the city itself, so that one must examine more closely its magnitude, characteristics and systems of absorption if one would predict its varying consequences.

Cairo has been attracting migrants from the rural areas at least since its transformation into a national metropolis at the time of Saladin in the twelfth century. Its wide fluctuations in population throughout the medieval period, however, suggest that some unknown but sizable proportion of Cairo's inhabitants were not true townsmen but rather what Claude Cahen has referred to as "floating population." Living in but not of the city, this population drifted cityward seasonally or periodically to flee oppressive taxation or to take advantage of expanding urban demand; it dispersed when epidemics or violence threatened to ravage the urban centers or when opportunities dictated a changed set of imperatives. Some migrants managed to remain in the city and become incorporated into the traditional craft and service structure. Given the consistent failure of

[15] We have computed, from data available in several tables included in *Annuaire Statistique 1960 et 1961*, a rank correlation coefficient between the average number of workers per establishment and the average weekly wages paid in Egypt in January 1959 for twenty-five industry groups for which this information could be assembled. The Spearman rank correlation coefficient (R_r) was +.771, indicating a close association between scale and remuneration. Inspection of the rank order of industries by size confirms our distinction between modern and traditional. The largest firms were in industries such as electricity and gas, water, sanitation, extraction of petroleum and processing its derivatives, etc. At the bottom of the list were such industries as wood working, extraction of stone and sand, leather processing and products, and foods and beverages.

urban populations to reproduce themselves and given the high urban mortality rates of this period, vacancies must have been created for these "floaters" to fill, even in a static hierarchy. During times of rapid growth, the proportion of migrants permanently remaining within the city must indeed have been impressive, since growth could have come from no other source.

While these early figures are but uncertain estimates, they gain some support from what appears, in the modern era at least, to be a remarkably stable ratio of migrants to Cairo-born in the demographic composition of the city. Throughout the recent period of reliable records, the proportion of Cairo residents actually born within the city has remained relatively constant at 60 to 65 percent. Migration, then, has been a permanent feature of Cairo urbanism during the present century and much earlier as well (see Table 2).

Although the proportion has apparently remained almost constant, the implications for the city have not. There have been important changes both in the magnitude and characteristics of the migrants and in the community into which they have been absorbed. First, there is the point of sheer numbers. In 1917, when the foreign-born constituted so large a minority, Cairo residents of rural Egyptian origin numbered fewer than 200,000. By 1947 there were more than 600,000 rural-born Egyptians living in Cairo; by 1960 there were over 1,000,000. It is reasonable to assume that the larger dimensions of today's city and the greater numerical strength of the migrants might not only permit greater organization along village lines for those wishing it, but also allow greater anonymity for those seeking to sever village ties. Migrants today probably follow more varied paths to adjustment than in the past. They also bring greater diversity with them.

Traditionally, migration in Egypt has been primarily a movement of young single males out of the villages. An unbalanced sex ratio—a male excess in the cities and a male deficit in rural areas—is an indirect measure of this selective migration. Extremely high sex ratios in cities usually indicate temporary migration, i.e., a high turnover and many urban-rural cross-currents, such as have been found in Indian cities. When the ratio begins to balance, it usually indicates a more permanent "settling in," a relocation of families and a reduced turnover of migrants.

There is clear evidence that this normalization of migration has been occurring in Egypt and that a move to the capital now represents a more total commitment to change than it formerly did. In the Cairo of 1897 and again in 1907, there were 112 males for every

100 females. This number dropped to 105 by 1917 but, under the impact of war-induced mobility, it rose again to 110 by 1927. Since then the ratio has been dropping towards normalcy, down to 106 in 1937 and 104 in the two most recent census dates.

TABLE 2

Population Enumerated in Cairo in the Censuses of 1917, 1947 and 1960 by Place of Birth.

Place of Birth	Enumerated in Cairo in Census of					
	1917		*1947*		*1960*	
	Total Population					
	790,939		2,090,654		3,348,779	
	Number	*Percent*	*Number*	*Percent*	*Number*	*Percent*
Population with known birth-place	787,461	100	2,085,987	100	3,331,078	100
Born in Cairo	ca.512,490	65[a]	1,325,485	64	2,079,434	62
Born outside Egypt	76,658	10	59,009	3	57,378[b]	1.6
Migrants from Egypt itself	ca.198,313	25[a]	701,493	34	1,194,266	36
Comprising:						
From other Governorates[c]	(22,293)	(3)	(58,998)	(3)	(102,587)	(3)
From Upper Egyptian Provinces	(ca.94,976)	(12)	(247,504)	(12)	(483,491)	(15)
From Lower Egyptian Provinces	(81,044)	(10)	(394,991)	(19)	(608,188)	(18)

Source: Selected tables in the Census of Egypt 1917 and of Cairo (special volumes in 1947 and 1960). (Computation ours.)

[a] In 1917 the boundaries of Cairo were enlarged to encompass the western bank of the river, an area which had formerly been within the Upper Egyptian province of Giza. Therefore, a sizeable but unknown proportion of the persons listed in the Census of 1917 as residing in Cairo but having been born in Giza province actually were not migrants at all but "victims" of a reclassification. [We have attempted to correct for this by subtracting some 29,000 migrants from Giza and adding them to the category. With this correction, the percentage of non–migrant residents declines to 61.] Logically, a similar correction should be made to compensate for Cairo's redistricting for the 1960 census; this correction would require a more complex operation and detailed figures for each of the census tracts added to the city, which unfortunately I do not now have.

[b] Includes a Syrian-born contingent temporarily in Cairo as a result of the political merger between Egypt and Syria under the then-common name, United Arab Republic.

[c] Primarily but not solely urban.

An even more sensitive measure of the nature of migration is the sex ratio of the migrants themselves. Unfortunately, this breakdown is not available for earlier periods, but in the most recent interval between 1947 and 1960 one notes a dramatic change in the composition of the migrant group. At the earlier date, only several years after the massive influx associated with World War II, the sex ratio of persons residing in Cairo but born in the provinces of Lower Egypt (the Delta) was 112, suggesting that at least some of the males who had migrated singly were already being joined by families. Since 1947, this trend not only has continued but also has been supplemented by a new form of migration—that of the nuclear family unit. We may surmise this fact from the 1960 sex ratio of Delta-born Cairo residents, which was an evenly matched 100.

The male temporarily separated from his family, however, was never as typical for Delta migrants as it was for those drawn to Cairo from the provinces of Upper Egypt, where not only economic necessity but also fear of the moral dangers of the city environment for women created a strong deterrent to female migration. Perhaps the most vivid evidence of this is found in the migrant sex ratio. In 1947, even after the major hump of war-related mobility began to level off, there were, in Cairo, some 405 males born in Upper Egypt for every 100 females. Since then the net inflow from the provinces of Upper Egypt has been heavily female. Between census dates the number of Upper Egyptian-born females living in Cairo increased by almost 162,000, which contributed much to a balancing of the abnormal sex composition. By 1960 the sex ratio for this group had already declined to 129 and, if we may safely project this trend, has by now approached an even greater balance (see Table 3).

Many of our theories concerning migrant adjustment must be revised in the light of this shift. The older observation that Delta migration is more permanent and acculturative than Upper Egyptian is probably no longer as valid. Both groups seem to be settling in more deeply. The current migration appears to be more adaptive and determined than either that of the floating populations of the premodern era or the transient male-dominated moves earlier in the century.

Changes in the structure of the city's economic opportunities are undoubtedly accelerating the process of migrant assimilation rather than hindering it, for where in the economy is there room for the newcomers? Certainly not in the rural sector which has begun to decline, not only proportionately but numerically as well. Farmland is being converted to industrial uses or into housing developments, and in the truck gardening periphery of Cairo many of the residents themselves are shifting to other forms of economic activity. Probably not in the traditional sector, for, at least within the older crafts, shifting demand is stranding superannuated proprietors and skilled workers in a backwater which at best can absorb only a tiny batch of ill-paid apprentices. In the nether world of services—some of dubious legitimacy, most of dubious productivity—there is room for the least prepared migrants, possibly thost who have flocked recently into the cemetery zones where the shock of culture adaptation is most softly cushioned. And domestic service still attracts a goodly number.

TABLE 3

Sex Composition and Sex Ratio of Egyptians Enumerated in Cairo in 1947 and 1960 but Born in Other Parts of Egypt, by Region of Birth.

	Population Born Outside but Residing in Cairo in . . .					
	1947			*1960*		
Region of Birth	*Males*	*Females*	*Sex Ratio*[a]	*Males*	*Females*	*Sex Ratio*
	— — thousands — —			*— — thousands — —*		
Upper Egypt	198	49	403	273	211	129
Lower Egypt	209	186	112	304	304	100

Sources: *Population Census of Egypt, 1947*, General Tables, (Cairo: Government Press, 1954), Table 11, pp. 60–61; United Arab Republic *1960 Census of Population*, Volume II, General Tables, (Cairo, 1963), Table 14, pp. 50–51.

[a] $\text{Sex Ratio} = \frac{\text{Males}}{\text{Females}} \times 100$

But the expanding opportunities are in the modern sector. Newcomers whose migration has been education- or army-linked move automatically into this sector, bypassing the traditional structure entirely.[16] Migrants with minimal literacy may qualify to learn the

[16] While most migrants who relocate in Cairo have been pressured out of their villages by land scarcity, economic setbacks or personal tensions, some have been drawn to the city as the result of by-product "uprooting". Education beyond the first few years often has required the student to commute to the nearest town. As a village boy rises through the school system he may be required to travel farther. University education means at least temporary relocation in the largest cities (Cairo, Alexandria or Asyut), from which the successful student usually does not willingly return. Some of this uprooting may be reduced by the present attempts to decentralize secondary education facilities. Service in the army appears to be a factor of increasing importance, on the other hand. Many a villager now in Cairo opted for urban life after an army discharge, having received some education and training in service which equipped him for opportunities most available in the metropolis.

skills which would in any case have to be taught to any new worker, even the one with the longest urban pedigree. For the unskilled and illiterate, the chance for a janitor's job in a new factory is probably better than his chance to sweep in the oldest workshop, since to obtain the latter position he would probably require an intermediary.

In short, today's migration has a complex impact on the nature of the city. The fact that it is family migration means, in some ways, a greater injection of ruralism into the city, for while the man may be acculturating, his wife may be recreating the physical and social accoutrements of village life. But on the other hand, the fact that the man is, in some instances, not only crossing the dimension of scale but also of time as well, and that his children are being educated to enter the expanding modern sector, means that migration may actually be stepping up the pace of change in the city, rather than holding it back by commitment to traditional urban ways.

With this general background, we can now turn to contemporary changing Cairo. We have tried to sketch the direction from which Cairo has come and some of the factors that are likely to influence her future course. It remains then to draw the contemporary cross-section.

VARIATIONS IN CONTEMPORARY CAIRO

The distribution of persons within a metropolitan community is never capricious. Subareas tend to specialize in accommodating populations with similar social characteristics and needs, levels of income, and styles of living. By studying the distribution of these characteristics, we can subdivide the larger, often confusingly diverse urban community into "subcities," each sharing a common geographic site, each housing a population which, although far from homogeneous, has certain general characteristics in common, and each playing its own unique role in the social and functional organization of the larger community.

While any casual observer realizes that districts within Cairo differ visibly (spanning a wider range of diversity than one finds in a western city) and enjoy reputations of greatly varying status and stereotype, the task of refining and measuring these differences, of delineating the boundaries of these subareas, and of including not only the more obvious sections of the city but also those virtually invisible nondescript zones that, in reality, constitute the largest expanses of any city, requires more than uncontrolled observation. By means of a scoring technique derived from computing various socio-economic and demographic indices from census tract data, factor analyzing the matrices of their ecological intercorrelations, and weighting and combining the multiple standardized indices for each census tract in accordance with the factor solution, we have been able to subdivide Cairo into thirteen contiguous and distinctly bounded subareas, distinguishable from one another in terms of the type of life led by most of their residents.[17] The physical disposition of these subcities is shown in Figure 1 and the populations included in each subarea of the city in 1947 and 1960 are presented in Table 4.

Before analyzing these results, however, we must first define a little more precisely what we mean by "style of life," for it is this factor that we shall use to investigate not only socio-economic levels within the city but also the variations in life-styles which this paper has referred to as the rural, the traditional urban and the modern. It so happens at this juncture of Egyptian history that these three types are correlated roughly with socio-economic status, but for any given individual the selection of the quarter of the city in which he lives is probably determined more by life style than by income, above a certain requisite minimum.

The operational definition of "style of life" is found in the factor loadings of our input variables upon the first and most important vector, Factor I. Variables of maximum significance are the proportion of men and women who are literate, the usual age at which women in the census tract tend to marry, the general level of fertility, and the degree of house overcrowding (ratio of persons per room). Secondary measures are religion, school enrollments, and the age of marriage for males. In many ways these are indirect estimates of even

[17] Detailed information on the selection of indices, their computation and statistical manipulation, is presented in "The Ecology of Cairo, Egypt," cited above. In brief, the method of factor analysis identifies mathematical vectors capable of accounting for the relationships and interdependencies observed among simpler variables. A factor is therefore a hypothetical "force" and is the measure of the variance *common* to several variables. Interpretation is required before this common vector, derived mathematically, can be given a name which is appropriate to the social reality to which it corresponds. On the basis of the original list of variables, replicated in 1947 and 1960, principal axes factor analysis yielded seven group factors of decreasing significance. Factor I, which we have named "style of life" was by far the most important, accounting for almost half of the entire variance in the correlation matrices of each separate year. It was therefore used, in a manner similar to that first suggested by Margaret Hagood et al., "An Examination of the Use of Factor Analysis in the Problem of Subregional Delineation," *Rural Sociology*, VI (1941), 216–33, to determine the major social divisions within Cairene population according to a unidimensional weighting system.

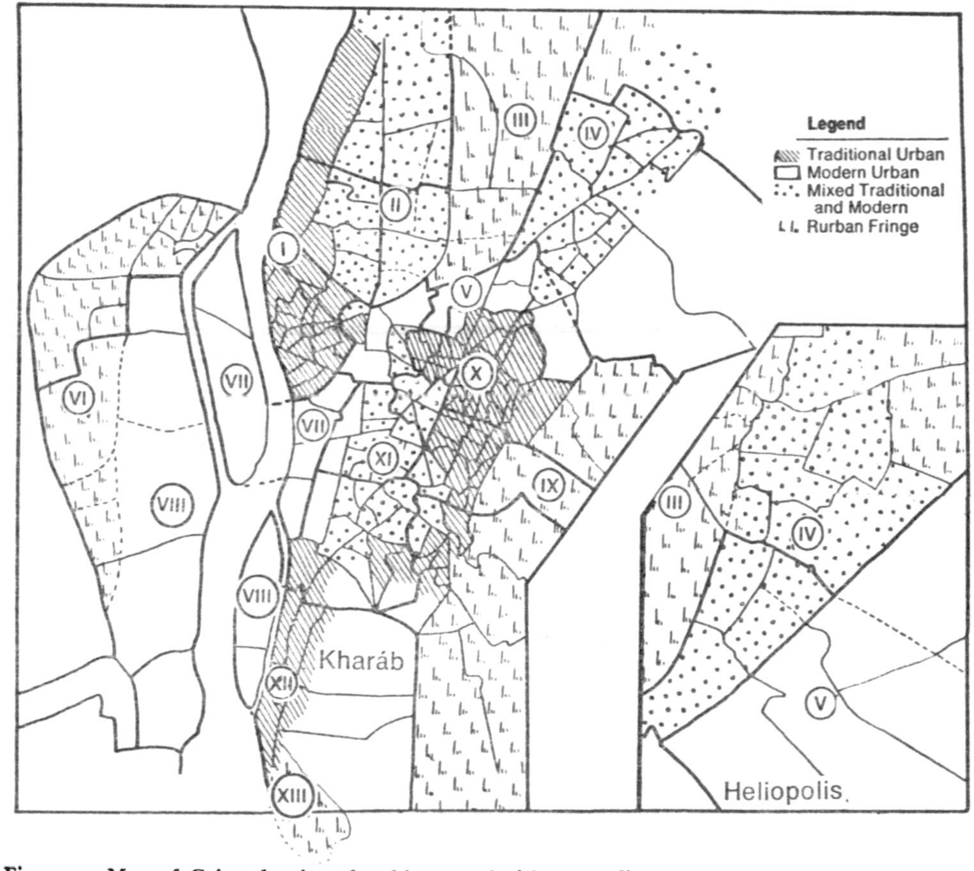

Figure 1: Map of Cairo showing the thirteen sub-cities according to style of life, 1947 and 1960.

more critical items—such as income, occupation, housing quality, place of birth, values and patterns of social relationships—for which we unfortunately have no detailed data by area.

This operational definition can be given connotative significance by a looser but more meaningful descriptive summary. Census tracts (and the composite subcities derived from their combination) with the highest "style of life" scores, are those portions of the city that are most easily identified as modern urban. In these sections, the resi-

TABLE 4

Number and Percent of Population in the "Style of Life" Subcities of Cairo in 1947 and 1960.

Dominant Life Style	*Population in thousands 1947*[b]	*Percent of Total*[c]	*Population in thousands 1960*[b]	*Percent of Total*
Rural				
Northern (III)[a]	100		200	
Western (VI)	60	11	150	14
Eastern (IX)	50		90	
Southern (XIII)	23		42	
Chiefly Traditional Urban				
Būlāq (I)	267		350	
Medieval City (X)	391	37	474	30
Miṣr al-Qadīma (XII)	103		188	
Mixed Traditional and Modern Urban				
Zaytūn-'Ayn Shams (IV)	187		456	
Shubra (II)	282	37	541	40
Transitional Belt (XI)	293		362	
Chiefly Modern				
Bāb al-Ḥadīd-Heliopolis (V)	172		261	
The Silver Coast (VIII)	55	14	220	16
The Gold Coast (VII)	65		82	

[a] Roman numerals after subcities correspond to map location.

[b] Total population as rounded is 2,048,000 in 1947. This is smaller than the official census total for the city because of the elimination of several census tracts and the rounding of other populations. Total population as rounded is 3,416,000 in 1960. This is smaller than the official census total because of our changes in boundary, our elimination of several census tracts and the rounding of other populations.

[c] The percentage totals less than 100 because of rounding.

dential structures are almost without exception "international urban" in design and use, the shops have glass-fronted display windows and fixed prices, and the inhabitants, except for domestic servants and the providers of local convenience services, dress in western clothes. Furthermore, most of the residents (males more than females) are not only literate but well educated, work at occupations within the modern sector of the urban economy, marry at later ages, often exercise control over the size of their families, send their children to school, and, with few exceptions, keep them there. As late as 1947, reflecting socio-economic and life-style differentiations rooted in the nineteenth century and even more in the later colonial period, a larger percentage of these residents than chance alone would have accounted for carried foreign passports or professed a religion other than Islam. Since the Revolution of 1952 and the sizeable exodus of foreigners in 1956, however, these zones have become more properly the domain of the indigenous upper and middle classes of the city.

Census tracts with the very lowest "style of life" scores, on the other hand, are almost exclusively rural in character and, indeed, are located chiefly at the periphery of built-up Cairo. (It is only the irregular shape of the metropolis that allows the northern agricultural wedge of Subcity III to dip so deeply into the core.) Within three of these urban fringe subcities are found practically all of the men still engaged in farming. In the fourth, the eastern-situated funeral quarters (the "Cities of the Dead," as they are commonly called), are found most of the even smaller number of workers serving as tomb custodians or engaged in quarrying and processing limestone and other minerals from the desert's rocky ledges at al-Jabal al-Aḥmar and al-Muqaṭṭam. Most men, however, are employed in diverse although largely unskilled capacities within the traditional sector of the urban economy. It is only their recent rural roots (either as migrants or ex-farmers) and their way of life that allow us to classify them as rural.

The housing in the three rurban areas is grouped chiefly in small village-like clusters and consists primarily of one and two story mud-plastered homes associated more with rural Egypt than with a metropolis. In the Cities of the Dead, inhabited tombs, village-style mudbrick huts, and multistory jerrybuilt structures of crudely fired brick cover the land more regularly. In all fringe areas commercial uses are minimal and confined to a narrow range of necessities, testimony to the low buying power of the residents who live near marginal subsistence levels. The women in these zones are almost all attired in the long black gowns, head kerchiefs and the supplemental shawls that adorn their country cousins; the men still wear the flowing *jalābiyya* during leisure, even those who may be required to don uniforms or trousers for work. Only a small percentage of the men can read and write and it is a rarer handful per hundred among the women who can do even this much. With little or no formal schooling to interfere with other life plans, men and women marry young. Most of the girls have married by sixteen and virtually all by twenty, the median age for marriage in the city as a whole. Early and sustained childbearing is their lot, as attested by the extremely high fertility ratios in these zones, even higher than in the villages of the hinterlands, thanks to the greater availability of medical facilities in the metropolis. Until recently the children rarely attended school. Even now, with the new compulsory education laws, the frequency with which girls are "overlooked" and the early ages at which most children disappear from the school system mean a very low rate of school enrollment.

Their way of life, however, is dying, since the land which sustains them is rapidly being converted to other uses. In the northern fringe a steady encroachment of modern industrial plants and of public housing projects preempts the land that was formerly held in large agricultural estates (*ʿizāb*) by the royal family and religious trusts. Subdivision lines on the drawing boards already intersect the fields on the western bank and two of the planned projects, Madīnat al-Muhandisīn and Madīnat al-Awqāf, are already being translated into streets and buildings. In the south, fingers of industrial development stretch up from Ḥalwān, and cement factories, thermo-electric plants and workers' compounds border alluvial fields that are still cultivated by *fallāḥīn* and irrigated by the water buffalo-powered *sāqiya*.

In 1960, some 14 percent of Cairo's population lived in these rural fringe areas, although not all of the inhabitants participated equally in rural-type activities. The most "urbane" of the four was the eastern cemetery area, until recently the abode of a long-standing urban population, the custodians of the tombs, shrines, and mosques, and the laborers in the limestone quarries and lime kilns—workers related to the two "natural resources" of the location. Since the intensification of residential densities in Cairo, however, this zone has become a reservoir for the overflow of both newcomers and the displaced poor, who find the openness of the Cities of the Dead preferable to the oppressive congestion of the "live" city.

In that same year, another 16 percent of Cairo's population lived in the three clearly modern zones of the city: the Gold Coast of Azbakiyya, Garden City, Qaṣr al-Nīl, and the Jazīra (Zamālik); the Silver Coast of riverine Giza, 'Ajūza, al-Manyal, and al-Duqqī; and the bulbous outlying suburb of Heliopolis together with the narrow band along the major northeast axis of Shāri' Ramses and the metro line that links it with the westernized business core at Azbakiyya.

Some 70 percent of Cairo's population, however, lived in districts where the style of life was neither predominantly modern nor still rural. Their homes were in the vast remaining stretches of the city in which traditionalism and modernism coexist uneasily. The blend, however, is in differing proportions and the amalgam itself is undergoing a daily transformation that helps us to predict the fate in store for these areas.

The heart of traditional urbanism still lies in the oldest remaining quarter of medieval Cairo, those portions of the Fāṭimid, Ayyūbid, and Mamlūk city that have not been "renewed" or whose former associations have persisted even after reconstruction. The irregularly shaped linear belt stretching from the Bāb al-Naṣr cemetery just beyond the northern wall down to the Citadel of Saladin, extruding westward through Bāb al-Sha'riyya almost into the lap of the westernized central business district, circumventing the newer developments at Ḥilmiyya (sited on the long since filled Pond of the Elephant, the Birkat al-Fīl), and encircling them on the south to encompass the area around the Mosque of Ibn Ṭūlūn, constitutes this city within a city. In 1917, the largest majority of all Cairene workers in the industries of textiles, dress and toilet, wood, leather and metal working —at that time almost totally traditional—lived and worked in this zone, as well as many of the proprietors, commercial dealers, and service workers associated with these trades. Today these enterprises are still located within this district, but they have been supplemented by modern plants and new products that have their locus far from the medieval core city.

Two additional quarters in the city have life styles that place them within the traditional sector. Not by chance do these happen to be contemporary outgrowths of medieval Cairo's two port suburbs: Būlāq and its northern extension along the Nile through the district known as Sāḥil (the shore), which inherited the port functions of Būlāq by the second half of the nineteenth century; and Miṣr al-Qadīma, the southern-situated port wedged between the Nile and the mounds that bury ruined Fusṭāṭ (the so-called *kharāb*).

Combined, these three subcities accounted for 37 percent of the total population in 1947 and 30 percent in 1960. While each has distinctive qualities that set it off from the remaining quarters and even from one another,[18] all have social and demographic characteristics with similar profiles. All are within the doomed traditional sector and all would be classified as slums by any outside observer. Yet they already show signs of having incorporated, if only in ungainly fashion, some elements of emergent modernism. Between 1947 and 1960 the Factor I scores in these subareas improved; these hidden changes are reflected in the more visible appearance of both the districts themselves and their inhabitants.

Within the three remaining subcities of Cairo, the modern and the traditional are more intimately interwoven and are moving toward a synthesis which foreshadows Cairo's future. Physically, each of these subareas mediates between and modulates the contrasting worlds that flank it. The sector subcity of Shubra, wedged between the traditional "*baladi*" world of Būlāq-Sāḥil to its west and the rural world of the northern agricultural fringe to its east; the elongated "workingman's" stretch of al-Qubba, Zaytūn, and Maṭariyya which separates this same agrarian zone from the modern urban quarter of Heliopolis; and the Transitional Belt which insulates and meshes the diametrically opposed urban worlds of modern Azbakiyya-Qaṣr al-Nīl-Garden City and the traditional core community around the ancient *qaṣaba*, are the crucibles in which the Cairo of tomorrow—maturing beyond the ethnic fissions and class extremes of yesterday—is being forged. Here typists and clerks, mechanics, electricians, and machine operators live side by side with and even within the same extended family households as petty proprietors, minor bureaucrats and simple workmen who follow older ways of making a living and more time-honored modes of social and economic involvement. Here too is to be found the urban life most typical of Cairo—earthy, noisy, gregarious, family-centered and ambitious for the younger generation.

The lower middle class is undergoing perhaps the most drastic social change of any group in the city, and yet the areas of its domain virtually escape notice, so overshadowed are they by the extremes of modernism and medievalism that tend to monopolize the observer's

[18] Būlāq is more highly industrialized, a development that dates back to Muḥammad 'Alī's selection of this zone for his experimental factories and technological institutes, and is more receptive to rural migrants. Miṣr al-Qadīma specializes in the pottery making, abattoirs and tanneries that were her heritage from the medieval period, and has retained, due to the presence of important religious edifices in the Qaṣr al-Shama', her attractiveness to Copts still within the traditional sector.

eye. If one wished to study the future of Cairo and, indeed, the future of Egypt itself, one could find no more crucial laboratory than these "grey areas" in which both rural and traditional roots are fast being exchanged for the future promised by the Revolution. The dreams being nurtured here, the pains that are being felt as the price of change, and the conflicts that are being resolved even now by the group most buffeted by the cross-currents and pressures alive in Cairo today, preview those which will beset more and more Egyptians as their society is transformed by modernization. These zones are transitional in more than a physical sense.

THE FUTURE AMALGAM

Thus we find in contemporary Cairo a complex microcosm of the forces at work in Egyptian society. The Middle Eastern city is not all of one piece; it is not simply a special "urban type" which differs from western cities by virtue of its unique Islamic heritage or by virtue of the particular culture in which it grows. Although these are important factors, they are not the only ones. Cairo contains within it the contrasting lifeways of the peasant village, the pre-industrial city and the modern metropolis. As such, it comprises a mosaic of subcities which exemplify each of these models. To the observer, these subcities present striking visual contrasts and seem at times to represent separate social worlds which coexist without interpenetration. Nevertheless, the most potent force now at work is the one leading to coalescence, to a blending of the extremes.

Only a few decades ago, the modern city was an insular anomaly, isolated from the traditional and rural subcities of the metropolis and having only tangential relevance to communal life. Worlds so insulated by social distance did not require physical barriers to maintain their separation. Mobility from one social world to the other was similarly restricted, obviating a demand for "transitional zones" to accommodate groups in social transit.

As indigenous forces toward modernization have grown, however, assimilation of the technologies and social techniques of "industrial society" into Egyptian culture and the absorption of a larger and larger proportion of Cairenes into the expanding network of the modern economy have tended to blur the lines between Cairo's separate social worlds. The transitional zones of the physical city, which reflect social worlds, have widened and spread. Extremes at both ends of the social spectrum are disappearing. The upper extreme is moderated by the exodus of the foreign communities, while the lower limit, leavened by compulsory education and consolidated in the new factories, is upgraded. The greatest growth is taking place in the interstitial zones where modern and traditional meet and where each exercises a molding and transformative influence on the other.[19] In the last analysis the inroads of the modern city upon the traditional can only deepen. Between 1947 and 1960 the range of social characteristics delimiting the traditional and modern poles of Cairo shifted upward, although improvement was far from consistent. (While literacy rates increased, so also did fertility ratios and the degree of room overcrowding.)[20] We can expect this trend to continue, as the older guardians of traditional ways are gradually replaced by a younger generation nurtured and educated in the less defensive atmosphere of post-colonialism. The earlier conflict between modern (which implied *faranjī*) and traditional (Egyptian) has essentially been resolved; it is possible to be modern *and* Egyptian.

Ruralism has little future in tomorrow's Cairo. At present, the gap between the urban quarters of the metropolis—both traditional and modern—and the residual and shrinking quarters of rural reserve, is substantial and growing. As improvements have taken place within the urbanized sections, the rural fringe has fallen relatively farther behind.[21] Furthermore, the expansion of the city will have to take place at the expense of these low density agrarian zones, which will gradually be displaced beyond the city limits. The rural enclaves near the center of the city have already met with this fate; the ones on the periphery cannot long avoid it.

Migration, however, will continue for some time to reinforce rural patterns within the city and to tie, through kinship and countermovements, the metropolis with the villages of Egypt. But even here, coalescence is the eventual although long-range probability, for the villages of Egypt cannot forever remain outside the mainstream of modernization. As rural improvements take place, the cultural and psychological distances between the rural points of origin and the

19 Figures supporting the increased "homogenization" of the city, the elimination of the "tail" of extreme upper scores, and the consolidation of the transitional districts can be found in "The Ecology of Cairo, Egypt," (cited in Note 3) chapter 5.

20 "The Ecology of Cairo, Egypt," chapter 4, especially Table 12, p. 202.

21 "The Ecology of Cairo, Egypt," Table 13, p. 288 with explanation in chapter 5.

urban points of destination of Egyptian migrants must decrease, so that, even though Cairo continues to attract new residents, the impact of the migration upon ways of life in the city will alter.

The final factor concerning further coalescence is an intangible one, but perhaps one of greatest significance. The kinship structure in the Middle East has always served to bridge divergent and divisive social groupings, linking classes with one another and city dwellers with their country cousins. This traditional force of cohesion is now being supplemented by a more modern one—easier communications fostered by the national government. Combined, they should further unify Egyptian society. And as divergent elements of that society coalesce, their coalescence should be reflected in the capital city as well.

POST SCRIPT

This analysis has been confined to only one contemporary community. This does not signify that Cairo is unique in the Islamic world, even though she may have stronger pulls to the land and the past, a more vigorous impulse to the future, and greater practical difficulties in adjusting simultaneously to growth and change on a massive scale. The analytical concepts and some of the conclusions in this study of Cairo have wider application throughout the Middle East. Practically all but the newest cities in the region contain a core of traditional structure, similar to Cairo's, which must be gradually integrated into a larger and coherent community. With few exceptions, all Middle Eastern cities have inherited a somewhat incongruous, if not alien, modern structure which must be transformed and indigenized if the cities are to generate their own special form of modern life. All face, for some time to come, the continuing task of absorbing the rural populations that constitute the bulk of the countries of which they are but a marginal but "growing edge." Each Middle Eastern city will, however, resolve these problems in somewhat different fashion and at varying rates. Thus, even in the process of this resolution, each will undoubtedly retain unique qualities that are the ever-renewing source of varieties of urban experience.

DISCUSSION

The problem of unity and division, and the basis of coalescence in Cairo occupied the conferees. Professor Geiser pointed out the importance of spiritual factors in the unification of such diverse organisms as cities. Professor Abu-Lughod agreed that "Islam works to overcome class cleavages, cultural cleavages, and to decrease social distance even among people who are objectively very dissimilar." Professor Fernea added that "we sometimes don't observe the ways in which common allegiance to Islam works as a unifying factor within diverse groups in the city. Michael Gilsenan, studying contemporary Sufi brotherhoods in Cairo, has indicated that the Sufi brotherhoods have a membership which includes not only working class people and people of lower economic groups, but also professionals—doctors, lawyers, engineers, judges, and army officers who live in the same neighborhood and belong to the same Sufi brotherhood."

Professor Safran, however, cautioned that there are limits on the extent to which factors such as religion serve to unify populations. "It is true that there is some 'Islamic' quality which colors all of Cairo, but it is hard to say precisely what this is. Besides the apparent unity there are also many underlying tensions and failures of populations to coalesce—to take only one example, the agony and schizophrenia expressed in many autobiographies."

Professor Fernea then opened discussion on the problem of migration in relation to the organization of city populations and the factors which tend to favor or inhibit the eventual assimilation of different groups to each other—the "coalescence" of populations into a common society. He remarked that one of the major issues to come out of comparing African cities south of the Sahara with those north of the Sahara is the role of the voluntary association, "which is such an important feature of central and west African urban life. Voluntary associations, which include dance groups to satirize other migrants within the urban area, and various kinds of economic investment organizations, are also found in the Carribean, for instance, among migrant groups in the cities. On the other hand, there are voluntary associations in the Nubian population that have only short periods of strength during which time they attempt to provide such things as education programs and certain banking or loan facilities for their members. However, these seem to dissolve very quickly because of disputes, and end up as title-holding organizations for burial grounds (which are nonetheless very important for migrants coming into an urban center). They become funeral societies. It seems generally true that stable social organizations are particularly hard to form in Middle Eastern society which tends to be fragmented. Extreme duress or outside change may cause a temporary coalescence of peoples, but rarely do such groupings become permanent corporations. This seems to be the case for migrant populations and the quarters of the city. Under duress they are temporarily able to form some kind of central organization, but when times are peaceful, organization rapidly falls off. Further, when one looks at the political offices that persist, they prove to be the ones which have no executive authority attached to them, but rather entail subordinate administrative activities. The *mukhṭar* (village chief), for example, is merely a census recorder and a tax collector—an agent for the central administration in dealing with the rural population. Irrigation controls show a similar situation. The reason Damascene irrigation has persisted so successfully is because it was placed outside the political realm and has always been outside political struggles. The people who run it are otherwise of no political consequence. The same is true of quarters. Not until the necessary functions of the quarter are separated from its political life do stable offices come into being. Once matters get into the arena of daily political struggle, they are exposed to the vicissitudes of the persistent factionalistic tendency which seems to be inherent in Muslim kinship patterns and culture in general.

"Even quarters composed of kin and ethnic groups are very fragile institutions, constantly threatened by any dispute which rises within them. Lines of cleavage, as the generational depths of such groups grow, are always present and always a threat to the cohesiveness of the community. Here, however, the larger community does serve to unite people. The greatest service an urban environment provides for these groups is to provide a judge—an intermediary outside of the community—to settle their disputes. For example, British administrators and archeologists who worked in Iraq report that disputes are brought to them by the people of the area. They are asked to be judges. The irrigation engineer in the village where I worked was constantly troubled by the fact that the villagers persisted in bringing to him disputes which had nothing to do with irrigation; they wanted him to judge all of their problems. It seems to me that this desire to push the resolution of problems outside of the immediate ethnic kin group is an important basis of social ties within cities and villages."

Professor Nader added that this form of conflict resolution is common in Middle Eastern villages: "In the area of the southern Biqā' valley where I worked for a time, people sometimes took their troubles to the city for resolution, but sometimes sought help in other villages."

Professor Lapidus remarked, on the basis of his study of medieval times, that the outside arbitrator who resolves conflicts forms a bond between groups, but the choice of arbitrator may be based on a prior sense of affiliation to a larger group: "In medieval cities, quarters often had a sense of affiliation to a larger religious community, and the judge was their normal arbitrator, not an *ad hoc* arbitrator. We should not overemphasize segmentation, and then find ourselves unable to understand what holds the society together. The segmental communities participate in larger social orders, often religious communities. Here again Islam became important as an ultimate basis of social cohesion."

Professor Fernea continued the discussion of migration into cities and its impact on the form of city society. While Professor Gulick stressed the persistence of rural habits and the maintenance of close ties with rural areas, and Prof. Abu-Lughod saw differences in Cairo depending on the origin and social background of the migrants, Professor Fernea noted Prof. Abu-Lughod's point that migrants tended to move into the modern sector of city life: "After all, the migrant to the city is a man who has given up agriculture and who stakes his subsistence on an entirely different form of income. He has to attach himself to one of the private or governmental bureaucracies which provide employment, and his whole style of life—his interests, his problems, and his associates—inevitably changes. Rather than dealing with people as totalities as he did in his personal interrelations in the village, at least in some areas of his occupational urban career, he is going to be dealing with people in particularized roles."

Professor Gulick, who discussed this problem at length in his paper, observed that the experience of migrants differs considerably: "Even within Tripoli, where I worked a couple of years ago, there are quite different patterns. Some migrant groups, as groups, feel themselves to be under pressure. Simple economic duress is one kind. The Alawites in Tripoli are a case in point. They live in a particular quarter, a place where respectable people do not go. They are wage laborers if they can find work. For another example, a small community of Maronites which originally moved to Tripoli to work as agricultural laborers are still known as humble people, and they still have a sense of village solidarity. In contrast, those who do well economically abandon as much as possible their kinship and rural ties. This reflects the great internal tensions in Middle Eastern kin groups, for many wish to evade the importunities of country cousins. They avoid the old *ḥāras*, quarters, or neighborhoods, and disperse."

Professor Nader thought that we have yet to understand fully the character of migration and its impact on the city: "What is the structure of the migrant's experience in the city? Who are the people he encounters—who constitute the significant others in his search for identity within this urban environment? How relevant, in fact, is his rural experience? Some studies show a tendency toward segregation among the people who came from the same rural area that the migrant comes from. Other studies suggest a more heterogeneous social experience."

Professor Gulick called attention to an interesting point in Prof. Abu-Lughod's paper about the construction of the Madinät al-Muhandisīn and Madināt al-Awqāf housing projects for engineers or other government employees. He remarked on parallel developments in Baghdad: "Housing projects are being developed at very reduced costs, because they are all on government-owned land, or what was formerly community land. The same term—*madīna*—is applied to ninety housing groups. All these are professional or occupational groups employed by the government. An interesting question thus arises. Will new kinds of neighborhoods, based on modern occupations, emerge in Middle Eastern cities? In Cairo there is a communications city as well as an engineers group, and in Baghdad a policeman's city and an officer's city. Is this a modern version of the older *ḥāra*?" Prof. Abu-Lughod added: "There are also housing projects associated with outlying industrial plants, in which a kind of syndicalist structure seems to be emerging—all the people working in a thermo-electric plant live in the housing unit set up for the electricity plant. Cement factory workers will be provided with housing near the factory. This is partly due to syndicalist philosophy, but also partly to a transportation system insufficient to move workers from the city to peripheral factories and back."

The meaning of such developments is not yet clear. Professor Fakhry detected an element of financial speculation as opposed to communal settlement: "The so-called *madīnas* in Egypt are limited to sixty or seventy houses, and many people have already sold out. They cannot be forced to keep the land and live there. Thus people from other professions move in." Professor Fernea noted the same thing in Iraq: "I remember that just before the revolution in Baghdad the Iraqi engineers had been given the opportunity to buy land in the Madīnat al-Muhandisīn, and over eight hundred engineers in Iraq, from all over the country, not just in Baghdad, invested money in plots of land with the expectation that even if they couldn't live there while serving as engineers, in twenty years, when they retired from government service, they would build a house in Baghdad. I knew several who refused to take jobs in the city because they felt they could save more money to this end while working in the countryside, and end up with houses in the suburbs."

As for general tendencies in Cairo's development, Professor Fakhry, referring to the persistence of rural population or of rural areas within or in the near vicinity of Cairo, remarked that some quite common economic factors are at work: "So much of Cairo is new, having been built in our own lifetimes, that people still hold on to and cultivate their land. Some people are obstinate and refuse to move or sell out. Others wait for higher prices. Thus rural population remains in the city." Concerning the general growth of Cairo, Professor Fakhry saw the economic factor as decisive. "Cairo is a booming city. Many people were attracted to Cairo. There was a time, especially during the two wars, when people suffered a great deal, and the only refuge was to go to the city. Then, as we all know, ambitious young people, or people who lost their property, or others obliged by social pressures to leave their villages came to Cairo. Another factor is the university education. When students came to Cairo, many families moved to the city in order to take care of their sons, and especially their daughters. There are many other factors. Cairo is an old and complex city, going through an experiment in history."

ROBERT McC. ADAMS

Conclusion

Professor Oleg Grabar spoke of the interdisciplinary character of Middle Eastern urban studies as "almost frightening." The accuracy of this description is well attested by the range and diversity of the foregoing papers. An attempt genuinely to "summarize" them accordingly would be not only foolhardy but futile. On the other hand, it may be useful briefly to outline some of the common themes running through them, as well as other themes that perhaps deserve to be more systematically dealt with. By concluding with an emphasis on these commonly included or omitted features, our discussions here should encourage a fuller interchange between specialists, and broader formulations of research strategy in the future.

To begin with, similarities or continuities between Babylonian and Islamic cities on the whole seem more impressive than their disjunctive features. There are repeated allusions in these papers to Islamic cities as social aggregations or composites, sharply differentiated along ethnic, religious, social, or professional lines, and taking the physical form of walled, somewhat hostile and independent quarters. This question was not explicitly dealt with by Professor Oppenheim, but it is worth pointing out that Sippar too was a composite settlement. Although the details are obscure, there were a series of separately named locales within the larger urban complex around which particular economic activities and tribal units apparently were grouped. Nor is Sippar alone in this respect. By the early dynastic period Lagash was already an equally distinctive geographic and social amalgam, while the existence of walled urban quarters is attested to archeologically at Khafajah. And Oppenheim notes the existence not merely of an upper-level administrative structure but also of a lower-level one, with the latter consisting of territorially localized "wards." Such wards seem indistinguishable to me in their basic structure from Professor Lapidus' "quarters," which he describes as "village-like communities within the urban whole" that were headed by appointed *shaykhs* and were often composed of recent village or Bedouin immigrants.

There were other important similarities between Babylonian and Islamic cities. The presence of concentric town plans was noted by Professor Oppenheim and Professor Issawi. Relative to pre-modern patterns elsewhere, it is clear that the Middle East has been characterized since very early periods by the unusually high proportion of its population residing in urban centers. Another continuing feature has been the relatively weak development of unifying institutions within the fabric of urban society, except in its uppermost strata. Below the level of supervening military rule, for example, Lapidus describes the Sufi orders and the *'ulamā'* as the only major exceptions to a general absence of institutionalized linkages between the lesser administrative and residential units. And the *'ulamā'*, he goes on to note, by the mid-eleventh century has substantially merged with landowning, bureaucratic and merchant families to become a fluid, long-lived urban elite of "notables." Leaving aside their special relationship to the propagation of an unprecedented body of religious law, this is surely very similar to Oppenheim's sketch of a closely intermarrying pool of judges, merchants, and scribes in Sippar who seem to have served almost interchangeably as officials and witnesses. Below the level of these Babylonian notables, it is difficult to perceive any evidence for permanent features of urban administration at all.

How may we account for this considerable continuity of settlement structure? The problem is underlined by the long intervening period of profound Hellenistic influence during which not all of the same conditions obtained; the rectangular grid pattern of at least the major Greek-founded towns is a case in point.Perhaps the answer is to be found in the same complex of factors very succinctly set forward by Issawi to account for the high proportion of urban settlement at the threshold of the modernizing period: the absence of a landed feudal nobility, any opposition to which would, as Max Weber showed, tend to impose greater urban unity; prevailing insecurity in the countryside; relatively favorable governmental treatment of townsmen as compared with rural peasantry; and important pilgrim or transit traffic.

The prevailing insecurity is particularly indicated by the high population flux not only in the countryside but also within cities. It would appear that Middle Eastern cities were characteristically unstable, as Professor Abu-Lughod suggested specifically about medieval Cairo, with substantial seasonal drift of population, sudden dispersions, and occasional periods of rapid growth. The persistence of similar phenomena until the present day is briefly noted by Professor Gulick and confirmed by other sources. Under such circumstances, a highly segmented urban structure may be almost inevitable. From an ecological viewpoint, and particularly if we regard city and villages as parts of an interacting social system, it may even be positively adaptive in the sense of tending to promote the survival of the population as a whole in the face of drastically changing conditions.

There are of course other features which seem to sharply distinguish Babylonian from Islamic towns, at least if we accept the testimony of these essays without regard for the differing contexts within which each was prepared. But one purpose of symposia is to scrutinize such accounts from a comparative perspective. To me at least, it would appear that some of the differences may arise more from superficial differences in the data than from differences in the basic structure we are seeking to apprehend.

Oppenheim, for example, provides a number of suggestive bits of evidence pointing to the corporate, semi-autonomous character of Babylonian cities. In this fashion he weakens if not undermines the old Weberian antithesis between Western and Oriental cities; yet Lapidus tells us that nothing of the kind was characteristic of Islamic urbanism. It is crucial to bear in mind, however, the contrasting nature of the documents on which Oppenheim and Lapidus had to rely. The former is largely dependent on contracts, letters, laws, and administrative records bearing "directly on the economy of the upper middle class, while officials, the military and the royal court are mentioned only incidentally." Is it surprising that accounts reflecting the activities and viewpoint of this group stress the corporate, autonomous qualities of urban life in Babylonia? Or that the largely religious, political, and genealogical accounts of the classical age of Islam should be unconcerned with or ignore vestiges which might have remained of such pre-Islamic urban features? Obviously we cannot posit the presence of features of Islamic urban society for which there is no available evidence, but at least we should be skeptical about the sharpness of the contrast under circumstances in which the character of the documentation is diametrically different.

A somewhat similar observation may be made with regard to the administration of the palace or temple. Oppenheim's sources permit him to view these entities in terms of their day-to-day functions leading him to describe both as "internal circulation organizations." Such a term seems entirely out of place within the usual picture of the medieval Islamic state, in which the rulers' own testimony repeatedly suggests that they were far more ruthless in the pursuit of autonomous goals for which the maintenance of local well-being was largely irrelevant. But admitting that this was undoubtedly true to a considerable degree, how much is actually known of the operation of Islamic landed estates, except in capital- and court-centered accounts of corruption and political deterioration? Very little, in my opinion. Yet it is only at the level of routine economic operations that the significance of apparent contrasts with earlier institutional arrangements could be conclusively tested.

The distinction between urban and rural is the one which contrasts most sharply in the accounts presented here of Babylonian and Islamic cities. Oppenheim speaks of the alternation of pro- and anti-urbanization tendencies, with the hostility of rural peasantry and nomads leading time and again to the destruction of cities. Lapidus, Gulick, and Abu-Lughod forcefully attack rural-urban polarity as a theoretical conception, citing much substantive evidence for the existence within urban centers of major population strata having more in common with the rural communities from which they came (for the most part, under varying forms of duress) than with the urban elites. But again it must be pointed out that Oppenheim's texts stem from that extremely limited part of the Babylonian population which was permanently committed to urban life, and it is by no means clear that their viewpoint was held in common by the urban population as a whole. There are, after all, numerous references in the same cuneiform sources to the forced resettling within city walls of deserting soldiers, corvée laborers, and other disaffected elements of the urban population whose own attitudes on the subject were never recorded for our benefit.

To a degree perhaps, this difference may arise from differential rates in the diffusion of contemporary ideas. Islamic urban studies, as already noted, are highly interdisciplinary in character; and even where they are not carried out directly under the banner of the social sciences, they are repeatedly crossed by caravans that are. Hence, the fact that formerly popular conceptions of a sharp rural-urban dichotomy have come under heavy attack from a number of related social science disciplines has become a commonplace. It may be suggested, admittedly somewhat cynically, that this impressively broad change in orientation has not yet penetrated into the more isolated (or more stoutly defended) terrain of Assyriology.

The reverse side of this coin also must be mentioned. Some of the rural-urban blurring which Gulick and Abu-Lughod document is a consequence of modern conditions which have no precedent, and hence should not be uncritically relied upon as a continuation of earlier processes. Mass media of communication, the mechanization of transport, and the world marketing of Middle Eastern oil all are examples of the modern industrialization process which have had a decisive impact upon both the spatial and the social distinctions between cities and their hinterlands.

The concept of cities not as isolated organisms but as constituents in a wide ecosystem seems to me a central one toward which the symposium as a whole has moved. This is not to deny that there are disjunctive points along the rural-urban continuum, but rather to suggest that it is more important to chart patterns of interaction and interdependence along the whole length of this continuum than to describe formal differences. To illustrate the relevance of this view it may be useful to return briefly to one of the basic continuities in urban structure that was attested earlier. The conclusion that traditional Middle Eastern cities were relatively segmented, unstable, and acephalous is not meant to imply a denial that there were important functional links, shared values, and unifying symbol systems that distinguished those cities and allowed them to survive as creative cultural centers. But it does argue for placing a complementary emphasis on ties—naturally, seldom attested directly in documentary sources—running through and beyond town walls. A number of terms used in the discussion following the papers suggest approaches that will aid in this enterprise, including references by Professor Fernea to the study of the role of voluntary associations and of urban representatives in the countryside, and by Professor Nader to networks of informal and sporadic but highly effective communication embracing both cities and villages.

Similarly, the traditional view of the "quarters" has been from above them, seeking to define their position in and responsiveness to the political and administrative structure. Approached from that standpoint, their lack of functionally differentiated, formal leadership emerges as a baffling problem. Here a complementary view from a lower level may be helpful, one which stresses non- or even anti-urban patterns of social control such as have been widely described by anthropologists and identified as closed corporate peasant communities.

While I have dealt more extensively with Babylonian cities than their role in the symposium justifies, I cannot forbear mentioning one further contrast. For all of their distortions and lacunae, the documents from towns like Sippar permit us to speak of real human behavior, not merely normative patterns. Sippar had twenty goldsmiths and only one potter; very well, this can be identified in at least potentially quantitative terms as part of its distinctive emphasis on caravan commerce and on trade with nearby nomads beyond the frontiers of settlement. Its urban elite we can also identify as entrepreneurs in the full sense, rather than as frozen stereotypes in a particular profession. They speculate in grain production, hire field labor, hold office, and combine in long-distance trading ventures.

What was the range of variation, or for that matter the average, of landholding among their Islamic equivalents? How much real mobility was there in and out of the class of notables? How often did office holding, landowning, and mercantile activities coincide not in paradigmatic cases, but in the typical individual? Answers to questions like these are vital to a deeper understanding of Islamic urbanism, yet they do not seem to be available until the threshold of the modern period. The major exception is our knowledge of Fusṭāṭ. Professor Goitein's many fascinating observations on the Geniza documents permit conclusions roughly paralleling those possible from cuneiform documents and make the full publication of this material a matter of exceptional urgency and importance. On the other hand, the special character of these documents as archives of the Jewish community in a unique center of long-distance commerce may limit the generality of conclusions drawn from them.

Another respect in which our discussion of Islamic cities has been somewhat deficient concerns the relationship of urbanization and urban phenomena to the broader sociocultural fabric of which cities are only a component. To what extent, as Gideon Sjoberg and others have argued, does a narrow preoccupation with urbanism overlook the fact that this is largely a secondary consequence of the more general, underlying process of industrialization and technological development? This question was implicitly touched on by Professors Issawi and Abu-Lughod, in that their data seems to show the beginnings of substantial urban growth while technology, capital investment, and the organization of production remained essentially traditional. On the other hand, Professor Issawi also referred to evidence from Egypt clearly reflecting the important influence of canal, railroad, and port construction and the introduction of cash crops, such as cotton, that were intended for European markets. Broadening the concept of industrialization to include these trends, it is by no means clear that industrialization was not the crucial, underlying factor even if its appearance in the Middle East was delayed until a later time. But whatever one's position on this question, it illustrates the issues to which our discussions have been pertinent and which we have heretofore managed to ignore.

The essential point is not that we bear any special responsibility for disentangling the complex problems of historical causation, but rather that we should continue to be sensitive to the broader contexts within which urbanization takes place. In recent times these contexts clearly include the expanding trade and omnipresent political influences of the European powers, and Issawi very properly points out that many specific features of Middle Eastern urbanization would be impossible to understand without them. Also included are the somewhat unexpected technological components that Abu-Lughod outlines, such as the very substantial expansion in the absolute numbers and importance of traditionally organized firms until relatively recently in the development of modern Cairo. One could only wish that more had been done earlier to record the character of traditional bazaar technology and craft organization, now rapidly disappearing.

Having urged the need for broader contexts within which Middle Eastern urbanism can be better understood, it deserves to be noted that we have taken very little cognizance of the worldwide aspects of the phenomenon. The substantial differences between Western European and Middle Eastern urbanization which Issawi describes do not clearly distinguish our region from the rest of the underdeveloped world. Are we prepared to countenance no useful modes of analyses that are cross-cultural and comparative? No studies in terms of general ecological, functional or locational hypotheses that will help us to perceive the organization of significant realms of behavior? Nothing in our region that can be called a "culture of poverty?" No personality types that are ubiquitous in urban situations? No common or converging trends in social structures? I hope we are not so parochial as to believe any of these things. But it must be admitted that our papers and discussion have given little evidence that we are prepared to take these wider trends and similarities into account even when seeking to understand the Middle East as a region in interdisciplinary terms.

My final point is closely related to the previous one. It can be introduced by a quotation from the same recent overview of urbanization to which Goitein has already referred:

> There is agreement, of course, that all cities bear certain resemblances to each other in both landscape and function, and that "systems" of cities have developed in all countries, evolving out of the socioeconomic conditions that characterize them. The controversial issue, one that intrigues geographer, sociologist, and historian alike, turns to a considerable degree on the relationship between value systems and social organization, on the one hand, and the development of city systems and various types

> of urban morphological patterns, on the other. It also involves levels of living and rates of economic development as they influence the nature of cities in various societies and countries. In other words, if types of urban hierarchy or urban morphology are taken as "dependent variables," to what extent is "culture" as an "independent variable" significant in "explaining" the differentiation among them?[1]

The tendency among Middle Eastern specialists, as this symposium again has shown, is to take cultural tradition as the *essential* independent variable in accounting for Islamic urbanism. To the extent that we have found convincing similarities with Babylonian urbanism, separated as it is by decisive cultural shifts and more than a millennium of time, there is perhaps reason to take into account some enduring ecological factors, although not necessarily to see Islamic urbanism in a wider comparative perspective. But culture is neither static through time nor constant as one moves from region to region within that immense congeries of diverse peoples and geographic settings that constitutes Islam. Hence we must expect—and indeed must look for—both temporal and regional differences in Islamic cities. Among the merits of the approach that Grabar and Issawi have taken here is that they point the way in this direction, even though "in the present state of our knowledge" Lapidus is surely right in insisting that Damascus and Aleppo still may serve as descriptions and models of classical Islamic city society throughout the Middle East.

The crucial point is one which Grabar has touched upon, but I would shift his emphasis slightly. He speaks of the understandings reached from references to the generalized Islamic "urban ideal" and from studies of particular urban centers, as partially contradicting one another. Recognizing that this is only a variant of a problem facing all historians, he considers how we might "bridge the gap" between the two extremes. I am skeptical as to whether there is anything to be gained by bridging that gap at all, and would suggest instead that it is precisely the tension between accepted historical generalizations at the regional scale and the understanding which emerges from the study of particular cities which will orient further research and make it both necessary and meaningful. In a quite analagous fashion, Grabar notes that the conclusions drawn from archeology often do not appear to coincide fully with the testimony of documentary

[1] Norton Ginsburg, "Urban Geography and 'non-Western' Areas," *The Study of Urbanization*, P. M. Hauser and R. F. Schnore, eds., (New York: Wiley, 1965), pp. 311–12.

sources, leading him to wonder whether dominant behavioral attitudinal patterns actually contrasted sharply with official statements, or were merely so obvious that they were seldom directly recorded. These, I submit, are precisely the kinds of questions about Middle Eastern urbanism on which it is now time to concentrate.

INDEX

Abadan, 108, 121
'Abbāsid empire, 22, 23, 48, 52, 59
Abū Bakr: Caliph, 31
Abū Muslim, 63
Achaemenid empire, 21
Acre, 141
Adab, 43
Adana, 110, 116
Aden, 108
Administration. *See* Bureaucracy; Cairo; Mesopotamia; Officials
'Aḍud al-Dawla, 63
Africa, 82, 162, 183
Agadir, 108
Agriculture, 74, 114, 115, 124; cash crops, 108, 124, 164; farmers reside in towns, 7, 13, 64, 65, 106, 111, 122, 124; estates, 177, 191; laborers, 13, 114, 185; in suburbs, 6, 64, 65; tenants, 13, 124; in towns, 64, 65, 86, 101, 106, 119. *See also* Cairo; Landlords; Peasants
'Ajūza, 178
Akhsīsa, 73
Akil tamkārē, 9
Akkadian, 4, 6, 16
Alawites, 149, 150, 185
Aleppo, 26, 70, 106, 118; citadel, 40; crafts, 110; law schools, 50–52; population, 102, 110, 116; religious buildings, 33, 38; social structure, 49; suburbs, 63; trade, 105, 107, 110.
Alexandria, 91, 96, 106, 113, 118, 147; education, 171; industry, 82, 117, 139, 166; literacy rate, 166; migrants, 83, 146, 150; mosques, 72; population, 102, 104, 108, 116, 152; trade, 82, 107, 110
Algeria, 108, 112
Algiers, 108, 116, 147
Alma Ata, 117
Ālum u šībûtum, 9
'Alw, 87
Amara, 149
Amasia, 110
Amīr, 91
Amman, 131, 132, 134, 145
Anatolia, 105, 146, 149; relations with Mesopotamia, 11, 16, 18
Andijan, 117
Ankara: population, 103, 110, 116; squatters, 146, 147, 149
Annaba (Bone), 108
Antioch, 104, 106, 107
Apprenticeship, 94, 171
Arab invasions, 57, 63
Arab world, 40, 42
Arabs, 5, 30, 57, 81, 141, 143
Aramaens, 18
Ardebil, 52
Artisans, 50, 67, 93, 94. *See also* Craftsmen
'Aṣabiyyāt, 51
Ascalon, 83
Aṣḥāb al-khabar, 91
Ashkabad, 117
al-'Askar, 61
Assemblies. *See* Sippar
Associations. *See* Fraternities; Migrants
Assur, 3, 5, 16
Assurbanipal, 7
Assyria, 7, 21
Aswan, 45
Asyut, 102, 117, 171
Ataturk, 117, 118
Athens, 5, 76, 106
Athil, 63
Attitudes toward cities. *See* Urban life: evaluation
Authority: Muslim concept of, 31
Autonomy: urban. *See* Communities; Empires
'Ayn Shams, 175
Ayyūbids, 90, 91, 178
Azbakiyya, 178, 179
Azerbaijan, 53, 55
al-Azhar, 36

Ba'ath, 156
Bāb al-Ḥadīd, 175
Bāb al-Naṣr, 178
Bāb al-Sha'riyya, 178

Bāb al-Shaykh, 127
Bāb al-Shaykh Bani Sa'īd, 127
Bāb al-Tibbānī, 149
Babtum, 9
Babylon, 3, 5, 7, 9, 14
Babylonia, 1, 3, 4, 12, 14, 15, 16, 21; cities compared with Islamic, 188–192
Baghdad, 75, 107, 118, 129, 133, 135; cabarets, 139; crafts, 110; dialects, 144; education, 118; gangs, 127, 128; housing, 131, 134, 138, 186; industry, 117, 148; kinship, 126; law school, 53; markets, 148; migration, 145–148; mosques, 36, 71; political factions, 156; population, 103, 105, 116, 127, 128, 145, 152; prostitution, 138, 139; quarters, 51, 63, 64, 149; religious buildings, 33, 38; salons, 141; squatters, 121, 126, 149, 157; town plan, 61, 138; trade, 107; villages incorporated, 151
Bahrain, 108
Balkh, 52, 67, 68
Bamm, 72
Banāna, 86
Banū Mazyad, 63
Banū Wā'il, 86
Basra, 34, 51, 107; mosques, 35, 72; population, 103, 108; quarters, 64; religious community, 55
Basta, 153
Bāṭiniyya, 91
Baybars' mosque, 40
Bayhaq, 68
Bazaars, 86, 87, 194. *See also Sūq*
Bedouins, 44, 49, 64, 105, 188. *See also* Nomads
Beirut, 107, 118, 119, 129; commuters, 150; dialects, 144; foreign residents, 121; households, 132; migrants, 145, 146; population, 108, 116; prostitution, 139; quarters, 153; trade, 110
Bejaya (Bougie), 108
Benghazi, 108
Bidonvilles, 121
Biqā', 132, 184
Birkat al-Fīl, 178
Bizerta, 108
Bukhara, 37, 52, 53, 64, 67, 68, 72
Būlāq, 163, 165, 178, 179
Bureaucracy, 22, 117, 120, 123, 185
Burhān, 53
Bursa, 103, 110, 116
Būrrī al-Lamāb, 151
Business. *See* Partnerships; Trade
Buwayhids, 63
Byzantium, 82, 83, 107

Caesarea, 82
Cairo, 24, 25, 26, 80, 83, 96, 159, 160, 194; administration, 91, 164; census, 163, 173, 175; churches, 85; citadel, 40, 61; Cities of the Dead, 176, 177; crafts, 110; economy, 164, 165, 166, 187; education, 118, 171; farmers, 164, 166, 176; foreign residents, 165, 176, 181; households, 131, 133, 134, 155; housing, 25, 85, 176; industry, 90, 117, 177, 178; labor force, 164, 165; literacy, 177; lower classes, 85; migrants, 91, 150, 159, 167–169, 170, 176; minarets, 39; modern sector, 176, 178, 181; mosques 33, 36, 37, 38, 72; occupational structure, 165; officials, 84; police, 91, 92; populace, 84, 105; population, 102, 105, 116, 152, 162, 168, 175; population characteristics, 172; population density, 177; population growth, 120, 146; property, 91; quarters, 63, 64, 85, 150, 176; ruins, 95, 178; rural sector, 101, 165, 166, 170, 171, 176, 177, 179, 180, 181, 187; slums, 179; styles of life, 160, 161, 166, 173, 176, 178–180; subcities, 172–180; suburbs, 63, 138, 163, 165, 169, 175, 178, 179; Sufis, 183; synagogues, 85; taxation, 91, 92; town plan, 40, 61, 86, 163; trade, 82, 107, 110; traditional city, 163, 178, 181; transitional zones, 178, 179, 180; truck gardening, 171; water supply, 93. *See also* Fusṭāṭ
Calah, 3, 5
Caliphate, 31, 32, 51, 71, 77, 84
Canals. *See* Irrigation
Caravansary, 51, 90
Carthage, 82, 106, 107
Casablanca, 108, 116
Census data, 163, 173, 175
Central Asia, 26, 116
Chimkent, 117
Christians, 22, 57, 80, 85, 107; guilds, 94; households, 132, 134; quarters, 51, 81; tax farmers, 84. *See also* Churches
Churches, 28, 30, 41, 85, 90
Citadels, 40, 61, 63, 79, 178
Citizenship, 1, 15, 25, 94, 95. *See also* Communities
Clergy, 21, 29, 43
Clientage, 49, 50, 51, 137, 138, 156

Cluster city, 121
Coffeehouses, 129, 140
Commerce. *See* Trade
Communists, 156
Communities, 34, 94, 111, 184; Cairo, 182; Dhahran, 121; integration of, 50; Jewish, 92; Lebanese, 157; Muslim, 39, 50, 60, 73; peasant, 192; political, 77, 78; religious, 24, 55, 107, 183, 185; village, 140; urban, 77, 78. *See also* Corporation; Law Schools
Constantine, 110
Constantinople, 104, 106, 107
Contracts, 4, 9, 15, 16, 83, 84, 95, 190
Copts, 156, 179
Cordova, 36, 37
Corinth, 106
Corporation: city as, 8, 9, 14, 15, 17, 21, 24, 184, 190
Corvée, 7, 13, 15, 191
Crafts, 51, 81, 82, 86, 108, 137, 164, 167, 194
Craftsman, 65, 84, 89, 105, 122, 137, 150. *See also* Apprenticeship; Artisans

Damascus, 81, 82, 105, 118, 155; agriculture, 106; crafts, 110; gangs, 56; irrigation, 184; law school, 50, 51, 52, 54; *madrasas*, 33, 40; migrants, 49, 83; mosques, 33, 34, 37; population, 102, 110, 116; quarters, 39, 64; religious buildings, 33, 38; social structure, 49; trade, 107, 110; villages, 67
Damietta, 65, 83, 102, 110
Danizköy, 151
Dār al-imāra, 36
Darb, 64
Dawra refinery, 151
Demography. *See* Population
Dhahran, 121
Dialects, 74, 76, 144, 155
Dihqāns, 68
Dinar, 89, 92
Dirhem, 89, 90
Diyala River Valley, 11
Diyarbakir, 110
al-Duqqī, 178
Dūr-Sarrukīn, 5
Dushanbe, 117

Economic planning, 120, 121
Education, 118, 144, 154, 171, 181
Egypt, 4, 67, 84, 88, 91, 159; agriculture, 114, 164; capital city, 81, 82; degree of urbanization, 104; economic planning, 120; factions, 156; houses, 176; industry, 167; law schools, 53, 72; migration to city, 150, 168–170; population, 102, 110, 112, 162; town plans, 62; trade, 82, 83; *'ulamā'*, 53; urban populations, 102–104, 109, 114, 153; urbanization rate, 109
Elders, 9, 21, 24
Elites, 53, 141, 144, 154, 189. *See also* Notables
Emar, 11
Empires: autonomy of cities, 10, 22, 77, 78, 107; dominate cities, 1, 21, 24, 31, 111
Entertainment: in cities, 129, 138, 139
Erzerum, 103, 117
Eskisehir, 116
Ethnic groups, 4, 18, 49, 51, 57, 136
Euphrates, 11, 63
Europe, 41, 99, 102, 106, 121, 123, 162, 164; communes, 47, 60; degree of urbanization, 104; investments, 108
Exports, 14, 82, 83, 108. *See also* Trade
Eyvāns, 38
Ezibtu, 15

Factionalism, 54, 56, 95, 100, 126, 184; and kinship, 127, 156; political, 125, 127, 156; sectarian, 128
Factories, 118, 160, 166, 167, 181
Fallāḥ, 74, 177
Family, 8, 21, 24, 49, 60; administrative roles, 10; businesses, 167; extended, 125, 131; migration, 148, 172; nuclear, 125, 131, 134; religious activities, 38; residence, 88. *See also* Kinship; Marriage
Fanā Khusraw, 63
Fārāb, 72
Farmers. *See* Agriculture
Fars, 53, 55
Fāṭimids, 36, 63, 85, 86, 90, 91, 178
Fez, 26, 77, 110, 118
Foreign aid, 115
Foreigners: in cities, 108, 120, 121, 127
Fortresses, 6, 62, 70. *See also* Citadels
Founding of cities. *See* Urbanization
France, 83
Fraternities, 43, 49, 50, 59, 60, 99
Frunze, 117
Fundūq, 51
Funerals, 140, 150
Fusṭāṭ, 24, 96, 193; distinct from Cairo, 61, 84, 163; Geniza, 80; houses, 85; migrants, 83; markets, 93; mosque,

34; police, 91; properties, 87; ruins, 89, 95, 178; taxes, 92; town plan, 25, 86; trade, 82, 107

Gangs, 127–129
Garden City, 178, 179
Garrisons, 15, 24
Gaza, 119
Gecekondu, 147, 149
Geniza, 24, 74, 80–86, 88–91, 93–96, 101, 193
Ghetto, 80, 81, 111
Ghūṭa, 67
Giza, 169, 178
Gold Coast, 175
Goldsmiths, 12, 81, 192
Governors, 21, 37, 66, 77
Grain, 13–15
Granaries, 14, 15
Greek, 47, 107
Guilds, 17, 18, 25, 43, 49, 93, 94
Gurgan, 52

al-Hādī, 63
Ḥadīth, 26, 29
Haifa, 108
Ḥalwān, 177
Hama, 34, 102, 116
Hamadan, 52, 110
Hammurapi, 8
Ḥanafī, 52–55, 71, 72
Ḥanbalī, 51–53, 54
Ḥāra, 64, 185, 186
Ḥarāsa, 92, 93
al-Ḥarbiyya, 61
Harmal, 17
Haydarkhāna, 127
Ḫazannum, 9
Ḥejaz, 30
Heliopolis, 175, 178, 179
Herat, 52, 63
Ḥikr, 92
al-Hilla, 63, 103
Ḥilmiyya, 178
Hindus, 18
Homs, 102, 155
Hormuz, 67
Hospitals, 90
Household size, 130–135, 151, 155, 156, 173
Housing: inherited, 89, 90, 96; quality, 176; as property, 15, 25, 81, 85; public, 177, 186; ruined, 88–90; tracts, 138; zoning, 152
Ḥusayn b. Ṭāhir, 63
Ḥuṣn, 70

Ibn 'Asākir, 33
Ibn Baṭṭūṭa, 95
Ibn Duqmāq, 87
Ibn Ḥawqal, 38
Ibn Hishām, 45
Ibn Khaldūn, 30–33, 142
Ibn Sa'īd, 84
Imām, 31, 43, 56, 67, 72
Imāmate. *See* Caliphate
India, 82, 89, 95, 109
Indonesians, 109
Industrialization, 114, 115, 118, 124, 125, 147, 193
Industry, 90, 115, 117, 121, 123, 177, 186. *See also* Factories
Inheritance, 88, 89, 96
Iran, 42, 58, 105; *'aṣabiyyāt*, 56; industry, 117, 121; insecurity, 106; land, 114; mosques, 37, 39, 40; oases, 68, 78; population, 103, 112; *'ulamā'*, 53; town plans, 62, 64, 68; trade, 11, 16; urbanization, 109; urban-rural ties, 151
Iraq: government, 135, 138, 184; households, 133; industry, 117; land, 114; law schools, 55; migrants, 145, 146, 149; population, 103, 112, 142; town plans, 62; urbanization, 104, 109, 142, 145
Irrigation, 4, 14, 75, 78, 114, 184
Isfahan, 107, 121; crafts, 110; industry, 117; law school, 54; minarets, 39; mosque, 37; population, 103, 109, 110, 116; quarters, 52, 64; religious buildings, 33; religous community, 55; town plan, 62
Iskenderun, 119
Islam, 39, 73, 77, 99, 183; conversion to, 23, 30, 57; history of, 22, 23; law of, 87, 89
Ismā'īl, 164
Ismā'īlīs, 54, 91
Ismailiya, 116
Israel, 112, 114, 121, 141
Istanbul, 26; education, 118; industry, 117; migration, 146, 149, 151; minorities, 113; population, 103, 110, 116; town plan, 61, 62, 119
Italians, 107, 109
Izmir, 108, 110, 113, 116, 117

al-Jabal al-Aḥmar, 176

Jabal Ansāriyya, 149
al-Jadida (Mazagan), 108
Jaffa, 108, 119
Jambil, 117
Jāmi', 33, 38, 73, 74, 78, 79; and community, 72; in *madīnas*, 70, 71; relation to *'ulamā'*, 70, 71
Jaxartes, 21, 48
al-Jazīra, 79
Jazīra (Zamālik), 178
Jerusalem, 74, 90, 96, 106; as sanctuary, 30; migrants, 83; mosque, 34; population, 102; quarters, 52; religious buildings, 33
Jews, 85, 107; attitudes to rural areas, 25, 96; community, 92, 94, 193; documents, 24, 80; households, 132; influence on Islam, 30; law, 87; migration, 109; officials, 84; quarters, 81, 86. *See also* Geniza; Synagogue
Jidda, 44, 108
Jinn, 140
Jordan, 112, 113, 114, 145
Judges, 31, 50, 56, 83, 184, 185

Kadar, 72
Kairouan, 36, 63, 82, 107, 110
Karaite, 85
Karāmiyya, 54
Karbala, 106, 127, 128
Karrada, 151
Karrada Sharqiyya, 127
Karum, 11
Kashan, 157
Kassites, 12
Kazakhstan, 117
Kazerun, 52, 64
Kazimiyya, 127, 128
Kazvin, 52, 63, 64, 103
Kenitra (Port Lyautey), 108
Kerman, 55, 72, 107, 124, 151
Kermanshah, 110
Khafajah, 188
Khānqā, 33
Khāns, 66
Khārijīs, 55, 72
Khartoum, 116, 151
Khawlān, 86
Khazar, 63
Khokand, 117
Khorsabad, 3
Khurasan, 66
Khurāsānabādh, 63
Khurramshahr, 108
Khuṭba, 29, 71, 73
Khuṭṭ, 64
Khwarezm, 67
Kinship, 126, 182, 184, 185; and industrialization, 125; in factions, 127, 156; and politics, 125, 126; and work, 125; in urban settings, 125, 126. *See also* Family
Kirkuk, 121
Kish, 16
Konia, 103, 107
Koran, 27–29, 50
Kufa, 34, 64, 107
Kūra, 96
Kuwait, 112

Laborers. *See* Workers
Lagash, 188
Landlords, 93, 114; social power, 76, 119; urban residents, 6, 74, 105, 141; village residents, 67, 68
Latakia, 119
Law schools, 42, 52, 53, 72; definition, 50, 51; historical evolution, 53; and mosques, 71, 72; relation to state, 51, 58, 59; as religious communities, 54, 58, 59; in rural areas, 54, 55
al-Lawt, 128
Lebanon, 126, 132, 134, 135, 150, 154, 157; agriculture, 114; population, 112; villages, 95
Libya, 108, 112, 120, 121
Libraries, 140
Literacy, 141, 154, 155, 166, 171, 176, 177
Liturgy, 10, 11
Ludd, 73

Madā'in, 61
Madīna, 67, 69, 70, 71, 73–76
Madīnat al-Awqāf, 177, 186
Madīnat al-Muhandisīn, 177, 186
Madīnat al-Salām, 61
Madrasa, 33, 38, 40, 56, 65, 78
Mahalla, 83, 96, 102
Maḥalla, 64, 65
al-Mahdī, 63
al-Maḥdiyya, 82
Mahra, 86
Maimonides, Abraham, 86
Maimonides, Moses, 82, 93
Mamlūks, 40, 45, 49, 51, 52, 54, 56, 59, 105, 178
al-Ma'mūn: vizier, 86, 91
al-Manṣūr, 63
Manṣūriyya, 63
al-Manyal, 178

Manzal, 70
al-Maqrīzī, 33, 34, 38
Maqṣūra, 35
Maqṭa', 82
Mari, 16
Marriage, 25, 83–85, 88, 140, 150, 173, 177
Marmara, 119
Maronites, 149, 150, 185
Maryut, 82
Mashad, 33, 103, 106, 110, 121
Mashhad, 32, 33
Masjid, 28, 29, 32, 35, 37, 38, 44
Masjid al-aqṣā, 28, 29
Masjid al-ḥarām, 28
Maslīaḥ, 84
Mass media, 136, 151, 154, 192
Maṭariyya, 179
Mausoleums, 38, 40
al-Māwardī, 32, 33, 70, 71
Maydan quarter, 139
Maydāns, 37
Mayor. *See* Officials; Sippar
Mazara, 91
Mecca, 29, 30, 106
Medina, 29, 30, 35, 45, 106
Mediterranean, 4, 11, 41, 107, 162
Meknes, 107, 110
Merchants, 11, 21, 50, 65, 93, 105, 122, 137, 193; foreign, 6; invest in land, 11; overseer of, 14; patronage of religion, 45; in rural areas, 56, 67, 84; settlements, 11, 51; social standing, 17
Mersin, 119
Merv, 37, 52, 63, 64, 67, 72
Mesopotamia, 5, 13, 15, 16, 18; administration, 6, 7, 10, 14, 188, 190; cities, 3, 4, 7, 16; fortifications, 6; irrigation, 4; law schools, 55; *'ulamā'*, 53; villages, 142, 143
Migrants, 135, 161; assimilation, 100, 121, 147–152, 157, 168, 170–172, 185; associations, 183, 184, 192; bedouins, 49, 142, 188; drain on rural population, 111, 118; families, 168–170; floating population, 65, 167; foreign origin, 108, 109; impact upon city, 182; males, 168–170; motives, 105, 106, 114, 146, 147, 171; personalities, 147, 148, 185; and population growth, 111, 120, 122, 145, 146; quarters, 65, 149; return to villages, 5, 24, 83, 105, 150, 151; settlements, 65, 126, 142, 148–150, 157; and social change, 172; village ties, 64–66, 149–151, 154, 181; work in cities, 140, 148, 151, 171, 172, 185. *See also* Squatter settlements
Miḥrāb, 35, 36
Mīkālīs, 53
Minarets, 35, 39, 40, 44
Minbar, 29, 70, 78
Minority groups, 113. *See also* Christians, Jews
Miṣr al-Qadīma, 163, 175, 178, 179
Misṭāḥ, 86
Mobility, 141, 193
Modernization, 138, 152–154, 157, 180, 181
Moghul Empire, 99
Mongols, 40
Monasteries, 38
Morocco, 78, 96, 108, 112
Mosque of 'Amr, 85
Mosque of Ibn Ṭūlūn, 36, 37, 178
Mosques, 26, 27, 36, 39, 44, 45, 50, 62, 66, 90, 140; administration of, 31; and Caliph, 31, 32; classical, 36; design, 35, 37, 38; earliest, 34; evolution, 23; Muslim theory of, 20, 30–34, 46, 70–72; and quarters, 31; relation to other structures, 36, 37; relation to ruler, 23, 36, 43. *See also Jāmi'*; *Masjid*
Mosul, 38, 103, 107, 116
Mount Lebanon, 149
Muḥammad, 29, 30, 31, 45, 50
Muḥammadiyya, 63
Muḥammad 'Alī, 164, 179
Mukhṭar, 184
Municipal self-government. *See* Empires
Munsif, 150
Muqābala: survey, 164
Muqaddam, 77, 83
al-Muqaṭṭam, 176
Murabba'a, 64
Muṣalla, 34, 35
Muslim cities: compared with European, 76, 77
Muslims, 80, 94, 96, 132, 156
Mu'tazila, 54

Nagid, 83, 86
Nāḥiyya, 55, 70
Najaf, 127, 128
Namangan, 117
Nasaf, 52
Nationalism, 78, 99, 156
Nepotism, 135

Nile River, 21, 45, 48, 67, 85, 142, 170, 178
Nineveh, 3, 5
Nippur, 3, 7, 18
Nishapur, 52–54, 63, 64, 67, 72, 107
Notables, 2, 24, 77, 78, 189, 193
Notaries, 56
Nomads, 18, 69; and agriculture, 123; chiefs, 65; invasions, 23; Mesopotamian, 16; moral character, 142, 143; protection against, 68; sedentarization, 11, 12, 64, 65; trade, 11, 193
North Africa, 26, 77, 107; 121; crafts, 110; ethnicity of population, 57; migration to cities, 146–148; population, 108, 116; ports, 108; trade, 110
Nubia, 44, 45, 155, 183

Oasis, 68, 78
Officials, 50; Jewish, 83, 84; mayor, 135; of quarters, 49, 77, 91; royal appointies, 14; temple, 18; village, 184. *See also* Governors; Sippar
Oil, 114, 120, 121, 192
Oran, 108
Oratories, 28, 38
Ottoman, 40, 48, 59, 99, 107, 126
Oxus, 73

Palaces, 1, 6, 13, 91
Palermo, 81, 91
Palestine, 67, 73, 82, 105, 109, 133, 145
Palestinean academy, 84
Palmyra, 155
Partnerships, 25, 81, 88, 89, 94, 95
Parthian empire, 21
Patronage of religion, 39, 43, 44, 45
Peasants, 18, 67, 78, 114, 141, 192
Performance criteria, 156, 157
Persian Gulf, 4, 121
Persians, 30
Pharmaceuticals, 81
Pilgrimage, 106, 114, 189
Pluralism, 136
Police, 53, 90, 92, 93, 94
Polis, 21, 22, 47, 60
Population: statistics on town size, 102–105, 108–116, 127, 128, 145, 152; urban growth of, 114, 120, 146, 152, 153, 189. *See also* Cairo; Urbanization
Port Said, 108, 110, 113, 155
Ports, 99, 106, 108, 110, 114, 178
Prayer, 29, 31, 34, 38, 71, 140. *See also Imām*
Priests, 21
Primate cities, 115–118, 120, 121, 152
Property, 8, 86, 87, 88, 89, 91, 93, 95, 96
Prostitution, 138, 139
Public health, 152

Qabul, 129
Qāḍī, 50, 56, 58, 66, 67, 91
al-Qāhira, 61, 63, 163
Qarāfa, 33
Qarya. See Villages
Qaṣaba, 179
Qaṣr al-Nīl, 178, 179
Qaṣr al-Shama', 179
al-Qaṭā'i', 61, 64
Qibla, 27, 30, 35
Qinnisrīn, 70
Quarters, 60, 62, 64, 81; administration of, 91, 184, 188; class differences, 87; as communities, 24, 43, 127, 184, 185, 192; conflicts between, 57; culturally distinct, 129, 153; as ethnic groups, 51, 86; gangs, 56; Mesopotamian, 6; migrants, 65, 148, 149, 150, 185; minorities, 81; modern, 111, 176, 186; mosques, 31, 43; in Muslim cities, 49–54, 64; newly founded, 63; notables of, 77; religious activities, 38; in town plan, 64; tribal, 11, 12, 64, 85, 149; walled, 64, 107, 108, 188. *See also* Suburbs
al-Qubba, 149, 179
al-Qurīyāt, 44

Rabaḍ, 64
Rabat, 108
Rāfiqa, 63
Ra'īs, 53
Rank-size rule, 115–118
Raqqa, 62, 63
Raqqāda, 63
Rashīdī, 63
Rayy, 52, 55, 63, 67, 107
Refugees, 145
Religion, 30, 33, 173, 176, 189
Rent, 88, 89, 90, 92, 95, 135
Responsum, 93
Ribāṭs, 33, 38, 66, 78
Rīf, 25, 82, 83, 84, 96
Roman, 21, 22, 27, 47, 82, 94, 106
Rosetta, 102, 110
Ruins, 87, 89, 95
al-Ruṣāfa, 61, 148
Rustāq, 55, 70

Safavid empire, 48, 59, 99
Safi, 108
Sahara, 183
Ṣāḥib rab', 91
Sāḥil, 178, 179
Saida, 107, 110
Sākākā, 44
Saladin, 61, 167, 178
Samarkand, 33, 52, 54, 117
Samarra, 36, 107
Sanctuaries, 4, 29, 30
Sanitation, 9, 93
Sargon of Akkad, 7
Ṣarīfa, 121, 149, 157
Sassanian empire, 21, 22, 78
Saudi Arabia, 44, 112, 114, 120
Saydiyya, 151
Schools, 141, 152, 154, 171, 173, 177. *See also* Education
Sea of Marmara, 151
Sects, 54, 55, 72, 156
Sedentarization. *See* Nomads
Seleucia, 104
Settlements. *See* Town plans
Sfax, 108
Sharecropping, 13, 141
Shādyākh, 63
Shāfi'is, 52–55, 71
Shargh, 72
Shari'a, 50
Shari' bayn al-qaṣrayn, 40
Shari' Ramses, 178
Shash, 68
Shaykh, 49, 149, 188
Shī'is, 51, 54, 55, 57, 127, 132, 156
Shiraz, 52, 63, 103, 117, 121
Shiyākhāt, 163
Shops, 12, 137, 149, 150
Shubra, 175, 179
Shurugiyya, 126, 149
Sicily, 81, 91
Sikka, 64
Silk, 82, 84
Silver Coast, 175, 178
Sippar, 8, 16, 192; administration, 188; agriculture, 1, 13; assemblies, 10, 14, 17, 21; citizens, 15, 17; family inheritance, 12; irrigation, 14; landowning, 13, 15, 17; mayor, 1, 10, 11, 17; merchants, 193; notables, 2, 189; officials, 6, 9, 10, 11, 15, 17, 189; quarters 9, 11, 12; rent, 95; ruins, 95; slave trade, 12, 13; taxation, 7; town plan, 188; trade, 5, 11, 12, 13, 14, 16, 193; wards, 9, 188
Skikda (Philippeville), 108
Slaves, 12, 13, 17, 51, 52
Slums, 155
Social services, 157
Socio-economic status, 173
Somalis, 109
Soviet Union, 116
Spain, 89
Spaniards, 109
Squatter settlements, 121, 134, 146, 147, 149, 157
Suburbs, 39, 62–65, 74, 138, 150
Sudan, 112, 116, 117, 133
Suez, 108
Suez Canal, 114
Sufis, 58, 59, 66, 183, 189
Sukne, 155
Sumerian, 4
Sunnism, 54, 55, 57, 72, 76, 132, 153
Sugar, 87, 90
Sūq, 64, 137, 138, 148
Suwayqa, 64
Synagogue, 28, 30, 80, 82, 85, 86, 90, 93, 96
Syria, 54, 58, 132, 149, 169; agriculture, 114; bedouin raids, 105; degree of urbanization, 104; law schools, 53, 72; population, 102, 103, 112; tribalism, 56; *'ulamā'*, 52; urbanization, 109

Tabaristan, 57
Tabriz, 103, 107, 109, 110, 116, 117, 121
Ṭāhirids, 63
Tangier, 108
Tanta, 102
Tashkent, 116
Ṭawwāfūn, 92
Taxation, 25, 34, 81, 105; collectors, 84, 90, 91, 184; of land, 92; in lieu of military service, 7; notables responsible, 10, 53; of peasants, 74; privileges in, 7, 24; of quarters, 49
Taybay, 155
Teachers, 58, 157
Technology, 194
Tehran, 107, 117; population, 103, 109, 110, 112, 116
Tel Aviv, 108
Temples: ancient, 1, 6, 7, 9, 12, 13, 17, 18, 24, 30
Telmun, 11
Tetuan, 108
Textiles, 8, 82, 84, 178
al-Thaghr, 82
Tigris, 4

Tinnis, 65
Town plans, 189; bazaars, 86, 87; central business district, 138, 139; composite towns, 61–64, 68, 69, 163, 188; influence of mosque, 39; grid pattern, 189; Mesopotamian types, 4; modern cities, 138; Muslim, 33, 37, 61–64; spaces and streets, 6, 40, 41
Trade: role in urbanization, 107, 108, 110, 113, 138, 152. *See also* Sippar
Traditional urban life, 153, 164–167, 171, 173, 175, 176, 178, 179, 180
Transoxania, 72
Transportation: role in urbanization, 114
Tripoli: Lebanon, 119, 125, 135, 138, 185; dialects, 144; factions, 128; households, 132, 134; quarters, 149; population, 102, 108, 116, 145; rural ties, 150
Tripoli: Libya, 108, 116
Tribes, 7, 11, 45, 85, 124, 125, 126, 188; loyalties, 56, 127; mosques, 34, 35; Nubian, 45; quarters, 85, 148, 149; revolt, 77; *shaykhs*, 149
Tujīb, 86, 90
Tunis, 107, 108, 116, 118, 131, 132
Tunisia, 78, 91, 108, 112
Turkey, 42, 113, 146, 149; agriculture, 114; industry, 117; population, 103, 110, 112, 116; urbanization, 109; villages, 133
Tyre, 82, 91

'Ulamā', 50–53, 55, 56, 59, 72, 140, 189
'Umar: Caliph, 34
Umayyad, 23
Umdurman, 116
United Arab Republic, 169
United Nations, 121
United States, 123, 130
Universities, 116, 117, 121, 171, 187
Ur, 3, 11
Urban life, evaluation: accepted, 7; as anonymous, 135–137; emotional bonds, 47, 75, 95; as sinful, 138, 139, 141. *See also* Urban-rural antagonism
Urbanization, 21, 99, 100, 111–114, 153, 154, 194; ancient Mesopotamia, 4, 5; concentration of population, 115–117, 145; degree of, 64, 104, 124, 142, 189; founding of Muslim cities, 22, 63, 78, 79, 81, 107; historical rhythms of, 78, 102, 119–121; population growth, 119, 121, 124; rate of, 109–111, 119; region-wide, 118; royal policy in, 5, 7, 78, 79
Urban-rural antagonism: condemnation of urban morality, 135–142; folk-urban stereotype, 100, 122; religions opposed, 57, 58; rural opposition to cities, 5, 55, 191; urban dislike of rural life, 25, 47, 82, 83, 96
Urban-rural life compared, 64–68, 74, 75, 101, 122, 154, 191, 192; factionalism, 128; households, 130, 156; kinship, 125, 126, 135; life style, 159, 176, 177; literacy, 155; religious practices, 140, 141; social values, 135–140; trade, 137, 138; women's roles, 130
Urban-rural interaction, 110, 111, 122, 142, 151, 152, 157, 190, 192; common religious communities, 24, 54–56, 77; elites, 141; factions, 125, 127; influence on hinterlands, 144, 157, 158. *See also* Migrants; Villages
Uruk, 3, 5

Villages, 4, 69; cloth manufacturing, 66, 68; compared with towns, 65, 66; culture, 74, 152; definition, 75; depopulation, 105; engulfed by city, 64, 151; factionalism, 49, 56, 126; farmers, 122; fortified, 66; hinterland of cities, 137; households, 131; merchants resident, 56, 67; migrants from, 65, 142, 146, 151, 188; mosques, 78, 79; mutual aid societies, 150; Nile Delta, 142; Nubian development, 44; in oases, 68; Palestine, 67, 133; relation to town, 54–56, 64, 68, 69, 73, 74, 75, 76, 149, 150, 154; religious functions, 66; shopkeepers, 137, 138; social structure, 67; size, 67; town life compared, 64–68, 74, 75; urbanites resident, 67, 83, 84, 118, 157, 158; villagers, 142, 143

Wahhābi, 128
Wa'lān, 86
Wālī, 91
Waqf, 44, 53, 59, 92, 177
Wards. *See* Quarters
Wāsiṭ, 52, 63
Wasta, 135
Wazīr, 31
Westernization. *See* Modernization
Women: division of labor, 129; migration, 169–171; segregation, 129, 130

Workers, 13, 18, 122, 139, 140, 147, 150, 171, 176–178, 183, 185

Yazd, 103, 107, 110

Zamm, 73
Zarība, 86
Zāwiya, 33
Zaytūn, 175, 179
Ziyād Ibn Abīhī, 35
Zoning, 86
Zoroastrian, 22, 57
Zu'ar, 49

www.ingramcontent.com/pod-product-compliance
Lightning Source LLC
Chambersburg PA
CBHW021706230426
43668CB00008B/745
9780520323797